How To Repair
Electrical Appliances

How To Repair
Electrical Appliances

Gershon J. Wheeler

Reston Publishing Company, Inc., *Reston, Virginia*

ISBN: 0-87909-342-0

Library of Congress Catalog Card Number: 72-76815
Printed in the United States of America.

ACKNOWLEDGEMENT

The author would like to thank the
TOASTMASTER DIVISION of McGraw-
Edison Company for supplying infor-
mation and drawings for this book.

Contents

Chapter 1

Fundamentals

ELECTRICITY

Most electrical appliances used in the home are very simple devices and should give many years of trouble-free service. Basically, there are two types: appliances that provide heat, such as toasters, electric irons, and percolators; and those that have a motor to provide motion, such as vacuum cleaners, fans; and washing machines. Some appliances combine both functions, such as electric clothes dryers, which have a motor to tumble the clothes and a heater to dry them. When an appliance fails to operate correctly, the fault may be mechanical or electrical. Determining the *cause* of failure is the biggest part of servicing. It is not necessary to be an electrical engineer to do this, but some knowledge of electrical fundamentals is necessary.

To get a better understanding of electricity, it is useful to compare it to the water system in a household. Water reaches the home through large pipes or mains and is under pressure so that it will be forced out of an open faucet which may be many feet higher than the water main. When a tap is opened, a current of water flows in the pipe from the main to the tap. The velocity of the current depends on the diameter of the pipe and also on its smoothness. In general, the larger the diameter and the smoother the pipe, the faster will be the flow of current. However, there is a limit determined by the pressure in the main. Electricity is quite analogous. The water main is replaced by the electric power lines which lead to the house. The electricity is under "pressure". This pressure is called electromotive force (EMF) and is measured in *volts.* When a suitable connection is made to an electrical appliance, current flows, and the speed of this current depends on factors akin to the smoothness and diameter of the water pipe. In general, the larger the diameter of the wire, the faster will be the electric current. The conductivity of the metal used to make the wire is similar to the smoothness of the water pipe. Some metals, such as copper and silver, are excellent conductors. These are said to have low *resistance,* and current flows through copper, for example, much faster than through iron. Nickel and tungsten are poorer conductors which tend to slow down the current. The velocity of electrical current is measured in *amperes.*

Let us now assume that a long garden hose is connected to a water faucet. The pressure may be 80 pounds per square inch. If the hose has an inside diameter of about an inch and is quite smooth on the inside, water will flow through it rapidly when the faucet is opened wide. However, if the hose is only one quarter inch in diameter and not smooth, the flow will be restricted. If a larger hose is connected to the open end of the quarter inch hose, the water will not move faster through this added hose, since the velocity of flow is limited by the smaller hose. In the same way, if electric current flows through several conductors sequentially, the velocity of the current is limited by the total resistance in the circuit.

In one sense the analogy between electricity and water breaks down completely. If a water faucet is opened, water pours out. But electricity does not pour out of a wall outlet or an empty lamp socket even when the switch is turned on. Thus, electricity or electric current flows only in a *complete circuit.* This is illustrated in Figs. 1-1 and 1-2. At a power station a difference in electrical potential (EMF or voltage) is generated. Two wires run from the power station to every house receiving electricity as shown in Fig. 1-1. These wires go through a distribution center in each home and then to the various electrical outlets in the home such as wall receptacles. In effect then, each wall receptacle has two terminations which are connected all the way back to the power station or voltage source, as shown in Fig. 1-2. When an appliance or lamp is plugged into the receptacle, the circuit is completed and current flows. This is illustrated in (a). The wiggly line labeled "appliance or lamp" is the standard electrical symbol for a resistance and is used here to represent the electrical resistance of the appliance. In (b), the receptacle has nothing plugged into it, and thus no current flows, since the circuit is incomplete. Similarly, in (c), an appliance is plugged into the receptacle, but its switch is in the *off* position. Again, the circuit is incomplete, and no current flows.

POWER
STATION

Fig. 1-1   Power distribution

Electricity

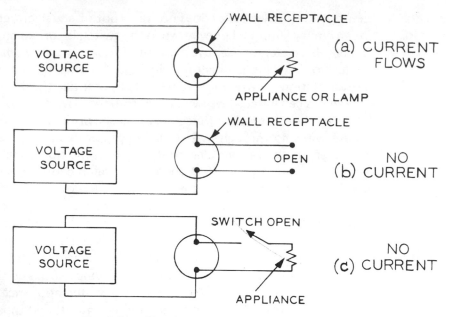

Fig. 1-2   Circuits

The difference in potential, or voltage, between the two terminals in a receptacle is what pushes electricity through a circuit. The two wires or terminals together form a *line,* and each wire by itself is referred to as a *side of the line.* In practice, for safety reasons (which are discussed in the next chapter) and for simplicity and economy of installation, one side of the line is always grounded. That is, it is physically attached to the earth. This side is then called the *grounded* side of the line, and the other side is the *hot* side. This means that a voltage exists between the hot side of a line and anything else that is physically attached to the earth. In a home, the cold water pipes are also *grounded,* that is, attached to earth, and thus a voltage exists between the hot side of the line and the cold water pipes. This is illustrated in Fig. 1-3. The plug is removed from the line cord leading to a socket containing a bulb, such as in a table lamp. One wire is attached to the hot side of a receptacle and the other to a faucet or water pipe. The bulb lights. If the wire at the receptacle is moved from the hot side to the ground side, the bulb will not light. Thus, the arrangement shown in Fig. 1-3 can also be used to determine which side of the line or receptacle is hot and which is grounded. Similarly, if you touch the hot side of a line and a ground, such as a water pipe, you will get a shock; but you will not get a shock from touching the ground side of the line and another grounded object. You will *not* get a shock even if you touch the hot side, as long as you do not complete the circuit by touching anything else that is grounded. This

GROUNDED SIDE

HOT SIDE

Fig. 1-3   Ground connection

means you must not be standing on ground or on a damp floor. This explains why birds are able to alight on high voltage wires without adverse effects. They do not complete an electric circuit with any part of their bodies.

If you look carefully at a wall receptacle, you will notice that one of the two slots for a plug is longer than the other. This is shown in Fig. 1-4. It is common practice to connect the longer slot to the grounded side of the line, as indicated in the figure. The screw in the center, holding

GROUNDED
SIDE

HOT
SIDE

Fig. 1-4   Wall receptacle

the mounting plate to the wall is also grounded in most installations. Although the slots are of different lengths, the plugs on most lamps and appliances have prongs which are the same width. Thus a plug can be inserted to connect either side of the appliance line to either side of the house line.

In industrial appliances, for safety reasons, it is desirable to attach the case of the apparatus firmly to ground. This is usually done by using a three-pronged plug and a special receptacle. The receptacle has a third hole which is round and connected directly to ground, as shown in Fig. 1-5. A wire from the case of the appliance is connected to the round prong on the plug so that the case is grounded when the appliance is plugged into the receptacle.

When electrical current flows through a resistance, it is converted to some other form of energy. This may be heat, light, or mechanical motion, or some combination of these. Thus, when current flows through the filament in an electric bulb, it is converted to light, and when current flows through the heating element in a toaster, it is converted to heat. However, the bulb also gets hot, and the heating element glows. Thus in both of these examples, and in every electrical apparatus, some of the electrical energy is converted to unwanted heat or light.

The unit of electrical current is the *ampere.* This is a unit of velocity indicating flow of electrical charges per second, just as water current is measured in feet per second.

Electricity   **6**

Fig. 1-5 Wall receptacle with ground connection

The amount of electrical *power* consumed by an appliance is measured in *watts*. For appliances which produce light or heat, the power consumed in watts is equal to the *product* of the *current* times the *voltage*. For example, one model of electric toaster is rated at 115 volts, 1320 watts. This means that in operation the toaster will draw 1320/115 = 11.5 amperes. Appliances which have motors that produce motion usually use less power than the product of current times voltage. For example, one model of juicer draws 1.75 amperes at 115 volts. The product of 1.75 times 115 is 201.25, but the juicer actually uses only 125 watts.

You pay for electrical energy on the basis of the amount of power used and the time it is used. The unit you pay for is the *kilowatt-hour* (Kwh). A kilowatt is 1000 watts and kilowatt-hours are the product of kilowatts and hours. Thus, if you used the 1320-watt toaster for one quarter hour and the 125-watt juicer for one fifth hour, you would use 1/4 x 1320 plus 1/5 x 125 or 330 + 25 = 355 watt-hours = 0.355 Kwh. Note that if a 50-watt bulb is left on for ten hours, it uses only half a Kwh.

When electric current flows through copper wire, there is some power dissipated, even though the wire is an excellent conductor with very low resistance. The greater the current for a fixed diameter of wire, the greater will be the power absorbed by the wire. This power loss becomes evident as heat. When too great a current flows through a wire, the heat generated may burn the insulation and cause a fire. The safe current handling capacity of a copper wire depends on the diameter of the wire. Thus, the choice of wire size, not only for house wiring, but also for line cords for individual appliances, is affected by the expected current.

The electrical system in a home consists of a service line from the main power lines to the house, a distribution box where the service line divides into many branch lines going to separate rooms, and finally the receptacles and sockets to which the electric appliances are connected. The usual voltage for most electrical appliances is nominally 115 volts, although they will work well at any value between 110 and 120 volts. However, some modern high-power appliances, such as electric ranges and electric clothes dryers require twice this value to furnish enough power to supply sufficient heat. (Recall that the power is the product of volts times amperes.) To supply both 115 volts and 230 volts, a three-wire service line connects the main power line to the house. This line, shown in Fig. 1-6, has one wire grounded, and each of the other wires is at a potential of 115 volts. Alternating current (AC) is used, and the voltages on the two hot lines are 180° or completely out of phase. This means that when one is 115 volts above ground, the other is 115 volts below, and the difference between them is 230 volts. This last voltage is used to operate a range, clothes dryer, and other high-

Fig. 1-6   Three-wire service line

voltage appliances, while small appliances are operated between either hot wire and ground.

Before considering the current requirements of house wiring, it is necessary to know what current is drawn by each appliance. Typical values of current for various appliances are shown in Table 1-1. Individual appliances may have slightly different requirements, but in general those appliances which supply much heat or do heavy work draw greater currents. A typical household may have 20 or 30 appliances, but fortunately only a few are used at any one time. The exact value of current drawn by an appliance or information from which this value can be determined is supplied either on nameplates affixed to the appliance, or from information engraved on the appliance itself. Some typical nameplates are shown in Fig. 1-7. Each nameplate carries at least the following information: the name and address of the manufacturer, the model number of the appliance, the voltage required (usually 110 to 120 volts), and either the power consumed or the current drawn or both. The symbol *UL* in a circle shows that the appliance meets requirements of the Underwriter's Laboratory. This is discussed in Chapter 2.

TABLE 1-1   TYPICAL CURRENT REQUIREMENTS

Appliances	Current Drawn (amps)
100-watt bulb	0.9
phonograph	0.5
TV set	1.6
refrigerator	2.0
small fan	0.5
blender	2.5
juicer	2.0
vacuum cleaner	6.0
freezer	5.0
garbage disposer	7.5
dishwasher	12.0
clothes dryer	20.0
toaster	12.0
iron	10.0
space heater	14.0
waffle iron	12.0

For appliances that supply only heat or light, the current in amperes is equal to the power in watts divided by the voltage. This is another way of saying that volts times amperes equals watts. Therefore, from the nameplates in Fig. 1-7, it is possible to determine that the iron draws 1100/120 or about nine amperes and the heater draws 1650/120 or about 13.8 amperes. The currents for the juicer and blender are specified at 1.75 and 2.5 amperes, respectively, because in appliances with motors the current drawn is uaually greater than watts divided by volts.

IRON

CAT.NO.47F6I. WATTS 1100. VOLTS 120. A.C. ONLY
GENERAL ELECTRIC CO.   BRIDGEPORT, CONN.
(UL) ALWAYS STORE IRON ON HEEL REST
            MADE IN U.S.A.

JUICER

MOD. NO. 02721
      PROCTOR-SILEX INC.
    CHICAGO, ILL. MADE IN U.S.A. (UL)
AC ONLY 110-120 VOLTS 60 CYCLES
1.75 AMPS   125 WATTS

BLENDER

          KENMORE
  MODEL 116-82421 (UL)
      110-120 VOLTS
  2.5   AC-DC   250
AMPS   WATTS
SEARS ROEBUCK AND CO.

HEATER

TOASTMASTER DIV. McGRAW-EDISON CO., BOONVILLE, MO.
MODEL 919 120 VOLTS 60 CYCLES A.C. 1650 WATTS (UL)

Fig. 1-7   Nameplates

The nameplate may have additional information as to proper use or care of the appliance. Thus, on the plate for the iron in Fig. 1-7, there is a direction for proper storage of the iron. This is to prevent scratching the face of the iron when it is not in use. Toasters usually have plates with additional information for two reasons. First, the

```
CAT. NO.  TO-5422
115 V.  1320 W.  A.C. ONLY   (UL)
WESTINGHOUSE
MANSFIELD, OHIO    MADE IN U.S.A.
PULL WALL PLUG BEFORE OPENING
```

```
            DISCONNECT TOASTER BEFORE
            OPENING TO CLEAN OUT CRUMBS
  (UL) 110-120 VOLTS  7.7 AMPS  50-60 CYCLES  A.C.ONLY
CAUTION: USE ONLY WITH WALL OR BASE RECEPTACLE
      MANUFACTURED BY TOASTMASTER DIV.
              McGRAW-EDISON CO.
              BOONVILLE, MO., U.S.A.
              MODEL NO. B160
```

Fig. 1-8   Nameplates for toasters

toaster is a *small* appliance which draws more current than many larger ones; and second, the exposed wires in a toaster are a potential shock hazard. Nameplates for two models of toaster are shown in Fig. 1-8. Note that both warn the user to "pull the plug" before opening the unit to clean out crumbs.

From the information in Table 1-1, it is possible to "design" the house wiring for a typical home. One possible arrangement is shown in Fig. 1-9. Each line in this figure represents a two- or three-wire electrical line. A three-wire service line connects the distribution box to the main power line. The service line is rated for 100 amperes and has a 100-ampere fuse or circuit breaker in series with it. The service line also passes through the electric meter before entering the box so that the amount of energy used may be determined. Several branch circuits emerge from the distribution box, and these are rated and fused according to their use. The *lighting* circuits are rated at 15 amperes each. These are wired throughout the house for lamps, small fans, electric razors, toothbrushes, radios, and TV sets. One branch line may service two or three rooms. A better arrangement is to have one branch for overhead

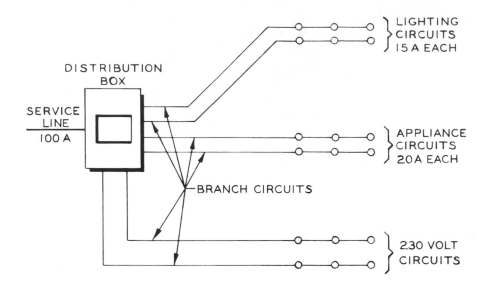

Fig. 1-9   Typical wiring arrangement

lights and one or more for wall outlets in living room and bedrooms. Since heavier loads are required in the kitchen, the *appliance* circuits leading there are rated at 20 amperes each. The overhead lights in the kitchen should be on one of the 15-ampere circuits. On the nameplate for the Toastmaster toaster in Fig. 1-8, the "Caution: Use with wall or base receptacle." is to ensure that the toaster will be plugged into a 20-ampere appliance circuit rather than a 15-ampere lighting circuit. The 230-volt branches are used for high-power equipment, and a separate branch should be used for each 230-volt appliance. These would include the clothes dryer, air conditioner, and range.

  If too many appliances are in operation on a single circuit simultaneously, the total current drawn can exceed the safe rating of the wire. To prevent overheating and possible fires from excessive current, each branch has a fuse or circuit-breaker which is designed to open the circuit when the rated current is exceeded. If an appliance has a short circuit, it will draw excessive current, since the short presents a lower resistance path across the line than the regular circuit. This will blow a fuse or open a circuit-breaker. However, an open circuit-breaker or blown fuse may also result when too many appliances are connected on one circuit so that the total current drawn exceeds the rating of the circuit. Overloading is not usually dangerous, but it can be a nuisance.

BASIC PRINCIPLES OF APPLIANCE REPAIRS

Electrical appliances should give many years of trouble-free service. In the first place, they are relatively simple mechanisms that suffer little wear and tear with proper use. Secondly, they are used relatively infrequently, in comparison, say, to an automobile. For example, a car that is used only two hours a day (a moderate amount of use) will log more than 700 hours of operation in a year. A toaster, on the other hand, may be used only five or six minutes a day, or about 60 hours a year. It is not surprising that toasters can give twenty or more years of excellent service.

Most faults in electrical appliances are caused by misuse or outright abuse. A piece of silverware dropped in a garbage disposer doesn't help either the utensil or the appliance. An electric iron may fall off an ironing board. An appliance is disconnected from a wall receptacle by yanking on the cord. A coffee percolator is left on after it is empty. This sort of act of comission or omission may not immediately cause an appliance to fail, but it does shorten the useful life of the appliance. If an appliance is operated carefully in accordance with instructions, it almost never needs servicing.

Nevertheless, for one reason or another an appliance may need repair, and you have to decide whether to throw it out, send it to a repair shop, or try to fix it yourself. If the appliance is still covered by a guarantee, you have no problem. *Let the guarantor repair it! This is important, since you may void a guarantee if you take apart a faulty appliance, even if you do nothing else to it.* After the guarantee period has expired, you may decide to fix the appliance yourself.

Should you do this? If you like puzzles, you should be able to fix a faulty appliance and may even find it fun.

First you must determine what is wrong. Frequently, the trouble is in the plug or line cord and it is not necessary to take the appliance apart. You should be sure that the fault cannot be corrected externally before you try to disassemble the appliance. If you know that it is necessary to take apart the apparatus, you have to figure out how to do this. Manufacturers don't like people to take apart their appliances, figuring that these people will do more harm than good, both to the appliance and themselves. Therefore, many appliances are apparently put together in some mysterious manner, and penetrating this mystery is part of the problem. Some of the disassembly techniques are discussed in later chapters, but in general it is best to consider the appliance from every angle before removing a single screw.

An appliance may not operate properly, but as long as it gives adequate performance, it is not necessary to repair it. Thus a toaster, for example, may toast bread unevenly, but as long as the toast is edible, its performance will probably be acceptable. As appliances age and their performance deteriorates, most users automatically "correct" for them. For example, if the temperature inside an oven becomes substantially different from that specified on the temperature control knob, a housewife usually becomes aware of the discrepancy and sets the control knob higher or lower as needed. Usually there is no complaint unless the appliance fails to work at all or blows a fuse when it is plugged in or turned on.

If an appliance doesn't work, that is, if nothing happens when the appliance is plugged in, it usually indicates an open circuit. This is shown in Fig. 1-10 as a broken line cord. Note that when the line cord is plugged into the receptacle, the toaster will not work since the electrical circuit is open. In this figure, the break is shown as being in the line cord where it is easily accessible and can be repaired quickly. However, the open circuit could also be inside the toaster. It could be caused by a simple disconnection or by a burnt-out element. Remember: the fault may not be in the appliance at all, but in the wall receptacle! *This should be checked first whenever an*

Fig. 1-10   Open circuit

appliance fails to operate completely. Plug in an appliance that you are sure is working.

If an appliance blows a fuse or trips a circuit-breaker when it is plugged in, the trouble is usually a short circuit, as shown in Fig. 1-11. Again, in the figure the trouble is indicated in the line cord. For example, the insulation between the two wires may break down from frequent flexing. The short circuit represents a path of very low resistance for electricity so that the current through it is very high. The excessive current trips the circuit-breaker in the line. The short circuit could also be located inside the appliance, in which case it would be necessary to take the appliance apart to locate the fault.

Fig. 1-11   Short circuit

It is also possible for an appliance to trip a circuit-breaker if it *overloads* the line. This happens when several appliances are operated on one line simultaneously, drawing current close to the rated value for that line. Now if an additional appliance is connected, the current drawn exceeds the rated value, and the circuit-breaker opens the circuit. However, this doesn't happen instantaneously as in the case of a short circuit, but instead the appliance works for a few seconds before the circuit-breaker is tripped. The obvious solution when a line is overloaded is to move some of the appliances to another line.

Some problems are purely mechanical. For example, a toaster pops the bread entirely out of the slots when it is toasted, or a dishwasher leaks because of a faulty gasket. In each case, replacing the defective part solves the problem. Again, the major problem in repairing may be disassembling the appliance to get at the defective part.

A more dangerous flaw occurs in some appliances when the wire running to the hot side of the line somehow makes contact with the case of the appliance. This is shown in Fig. 1-12. The defect can be caused by burnt or worn insulation or by a mechanical shock or vibration caused by dropping the appliance. It can sometimes be a defect in a new appliance which goes undetected for years,

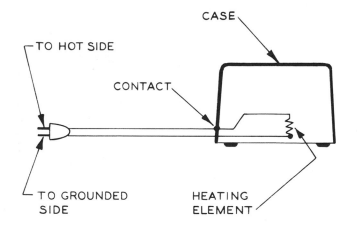

Fig. 1-12 Hot case

since the appliance still works perfectly even though the case is connected to the electric circuit. The potential danger of this flaw is that of electric shock, which can result if the case of the appliance and a cold water tap are touched simultaneously. A *temporary* solution is to pull out the plug, invert it, and return it to the wall receptacle. Then the wire touching the case will be connected to the grounded side of the line, and there will be no danger of electric shock. However, this condition should not be allowed to remain for long, and the flaw should be corrected as soon as possible. Incidentally, every *new* appliance should be checked for this flaw before it is put in use, since it is something that can escape factory inspection, and it may be lethal. This check is described later in this chapter and again in Chapter 2.

TOOLS

Most electrical appliances can be repaired with ordinary handtools. Although there are a few special tools designed to simplify repair of appliances, they are not worth buying unless you intend to go into the business. You should have an assortment of screwdrivers. Screws in electrical appliances come with a variety of different heads, shown in Fig. 1-13. Thus, you should have, besides the ordinary screwdrivers, a small Phillips screwdriver, an inexpensive set of Allen wrenches, and, desirable but not absolutely necessary, a set of nut-drivers which can also be used to tighten hex-head screws. A small crescent wrench is also handy if nut-drivers are not available. Wire strippers are usefull, but an ordinary pocket knife serves almost as well. You will need a pair of long-nose pliers to get into tight spaces, and some-

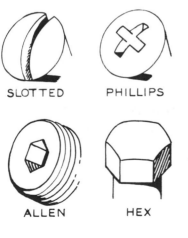

SLOTTED   PHILLIPS

ALLEN   HEX

Fig. 1-13 Screw heads

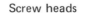

thing for cutting wire, such as diagonal cutters or side-cutting pliers. For larger appliances you may need a Stillson wrench, slip-joint pliers, and a hammer.

When using a screwdriver on a small appliance or on a sheet metal surface of a large appliance, you must be careful not to strip the thread. Threads in sheet metal, plastic, or even in soft aluminum and brass are easily stripped by exerting too much force on the screwdriver. This danger can be minimized by using the *smallest* screwdriver that will do the job. Screws should be tightened using a light finger pressure, never exerting torque from the wrist.

Since the electrical performance of an appliance must be checked before a diagnosis of fault can be made, some sort of diagnostic tool or instrument is required. Consider first the electrical checks you must make. You must find out if there is a short circuit or an open circuit in any part of the line cord or appliance. You also want to check to see that a voltage exists where it is needed and that there is no voltage where it may be dangerous (that is, on the case of the appliance). Note that accurate voltage readings or resistance values are not required. A volt-ohmmeter is helpful for making these measurements, but an inexpensive one will serve just as well as the most accurate, expensive models. If, however, you intend to service radios and TVs (not covered in this book), an accurate meter is necessary.

If you do not have a volt-ohmmeter and will repair only one or two appliances, you can make or buy a very inexpensive test-light which can be used to give you all the information you need to diagnose the trouble in a faulty appliance. This is simply an electric bulb in a socket with two wires connected to it, as shown in Fig. 1-14. The wires are insulated, and pointed probes are connected to the free ends. Alternatively, if you wish to make a test-light yourself, you can use stiff wire and leave about one inch bare at the free end of each wire. The stiff wire protruding from the insulation makes a satisfactory test probe. Since flexible leads are easier to work with than stiff wire, you may want to use flexible wire, in which

Fig. 1-14   Test-light

case you can stiffen the bare inch at the end by applying solder to it. You will also need a piece of wire two or three feet long with probes on both ends. One form of commercially available test-light uses a neon bulb instead of an incandescent bulb. It is very inexpensive and is a worthwhile addition to your tool kit.

Since the test-light described is used across the 115-volt line voltage, it must be handled carefully. Radio and auto supply stores also sell battery operated test-lights which can be used without danger of electric shock. However, before spending too much money on a test-light, you should compare the cost with that of a cheap volt-ohmmeter.

A few clip-leads will simplify servicing. These are short pieces of wire, about six to eighteen inches in length, with alligator clips at both ends. By clipping the alligator clips to various points in the circuit, it is possible to tie parts together electrically or to bypass parts of the circuit so that other parts may be checked alone.

BASIC TESTS

The primary test is a voltage check to make sure there is voltage where it is needed and no voltage where it may be dangerous. This check can be performed with a test-light as well as with a volt-ohmmeter, and indeed for the vast majority of troubles the test-light is adequate. Another important test is a continuity check to make sure the circuit is continuous. A test for a short circuit is a type of continuity check where you hope the circuit will not be continuous. A continuity check is performed easily using the resistance scales of a volt-ohmmeter. It can be done with a test-light but requires more care.

When an appliance doesn't work at all, the first thing to determine is whether there is voltage at the wall receptacle. This can be ascertained easily by plugging in a different appliance which is known to be in working condition or by checking the outlet with your volt-ohmmeter or test-light. If you use a meter, make sure that you set it for a scale higher than the expected voltage. Therefore, if you are measuring a 115-volt line voltage, use the 150-volt or 250-volt scales. If you use a low scale, for example, 10-volt scale, the 115 volts will knock the pin off scale and may damage the meter. The two leads from the meter are inserted in the receptacle as shown in Fig. 1-15, and the meter should indicate a voltage between 110 and 120 volts. The exact reading doesn't matter. Similarly, if you

Fig. 1-15 Voltage check with meter

use a test-light, the two leads are inserted into the receptacle in the same manner, as shown in Fig. 1-16. If the bulb lights, there is voltage at the receptacle. When inserting leads into a receptacle, make sure you hold them by their insulation, and never touch the bare wire! If it is difficult to do this, it may be easier to use one side of each half of the receptacle, as shown in Fig. 1-17. The bulb will light as before, since both halves of the receptacle are wired in the same way. If the bulb fails to light, the trouble may be in the receptacle or it may be a tripped circuit-breaker. Before tackling an appliance itself, make sure the receptacle is "hot".

Assuming the receptacle is hot, the appliance is now plugged in, but nothing happens. Voltage readings can be

Fig. 1-16 Voltage check
with test-light

made at several points inside the appliance using the test-light or the meter. In Fig. 1-18 several test points are indicated. If the two wires from the tester are connected to points A and D in the figure, a voltage reading would indicate that the cord is satisfactory. If now the top probe is moved to point B, holding the bottom one on D, failure to light the bulb or get a meter reading will indicate a faulty switch, which should then be replaced. On the other hand, if the switch is all right the next check might be between points C and D. A lack of voltage here would indicate a bad thermostat. Note that a thermostat is simply a switch operated by heat. If the thermostat checks out all right then the trouble could be in the heating element. In practice, a visual inspection might reveal the open circuit without the necessity of all the voltage checks, but the purpose here is to show that voltage can be measured anywhere in the circuit.

When a resistance is connected across the 115-volt line, there are 115 volts across the resistance, and this voltage causes current to flow through the resistance. If one end of the resistance is connected to ground, as shown in Fig. 1-19, the hot side of the resistance is 115 volts different from ground. There is said to be a *voltage drop*

Fig. 1-17 Alternative
method of checking receptacle

Fig. 1-18   Test points

across the resistance. If we were to connect one lead from our meter to a fixed ground connection and the other lead to the hot side of the resistance, we would read 115 volts on the meter. But if the second lead were connected to the ground side of the resistance instead of the hot side, we would read zero volts. If the lead were slid along the resistance from the hot side toward the ground side, keeping the other lead fastened to ground, the meter reading would begin at 115 volts and decrease gradually to zero as the lead was moved along the resistance. For example, in Fig. 1-19, if the hot lead to the meter contacted point A of the resistance and the other lead was grounded, the meter would read about 50 volts, since A

Fig. 1-19   Voltage drop across resistance

Basic Tests

is somewhat nearer the ground side than the hot side of the resistance.

The discussion in the preceding paragraph leads to a necessary voltage check on appliances. Assume the resistance in Fig. 1-19 is a heating element in a toaster. Assume also that because of carelessness in shipping, point A of the heating element has somehow made contact with the case. The toaster will still toast bread perfectly, but the case now has a voltage of 50 volts on it. If someone touched the case and a cold water faucet simultaneously, he would get a shock. This is a hazard in every appliance, and consequently every new appliance and every repaired appliance should be checked for this danger. To check, simply touch one lead from the meter to the water faucet and the other to the case of the appliance. The voltage should read zero. Notice that even if the voltage is only five volts, it indicates that something is wrong. Five volts will not give you a shock or even a tingle, but if someone pulls out the plug and then puts it back inverted, the ground side and the hot side will be interchanged. This means that the case, which before was five volts away from ground, is now five volts away from 115 volts. The resultant 110 volts is dangerous.

You can also use a test-light to make this check. Place one lead on the water faucet and one on the case. If there is 50 or 60 volts on the case, the light will glow at about half its normal brilliance. If there is substantially less, there might still be a hazard, but not enough voltage to light the bulb. However, if the plug is removed from the receptacle and inverted, there will be sufficient voltage to light the test-light. Thus, whenever you use the test-light, you should keep this possibility in mind and make checks with the plug both right side up and inverted.

If two points are connected by a short circuit, there should be zero voltage difference between them. A switch is supposed to be a short circuit when it is operating properly. However, every time a switch is opened or closed, a small electric arc jumps between the connecting points and tends to burn these contacts. After many switchings the contacts may be so burned or dirty that the switch is no longer a good short circuit, but presents some resistance to the current. There is then a voltage drop across the switch and thus less voltage across the appliance itself. The appliance might still operate but, for example, a toaster may take longer to toast bread, or a vacuum cleaner may have less suction. To check a switch, the meter leads are placed across it. Referring to Fig. 1-18, one lead from the meter is placed in contact with point A and the other with point B. With the switch open, you should read 115 volts, since point A is connected directly to one side of the house line through the line cord, and point B is connected to the other side through the appliance itself. When the switch is closed, the voltage should be zero. The meter scales should be changed downward, with the leads contacting points A and B, to increase the sensitivity. If no reading is obtained on a 5-volt or lower scale, the switch is probably satisfactory. Important: *always return the scale setting to something more than 120 volts before making further checks.*

A test-light like that illustrated in Fig. 1-14 cannot be used in this manner to check a switch, since the voltage across the switch may be insufficient to light the bulb. However, the inexpensive neon tester is sensitive enough to detect a very small voltage and can be used to check a switch.

Another type of test is a continuity check of the circuit or parts of the circuit. By means of this test you determine if the circuit is continuous or whether it has a break in it. The ohmmeter part of the volt-ohmmeter is used. The appliance to be tested is disconnected from the house line. Usually when you perform this test you want to know if the circuit is complete and you are not interested in exact values of resistance. The two leads from the ohmmeter are connected to two points in the circuit. If there is a break in the circuit, the needle on the meter will not move, but if there is continuity, the needle will swing to the right. When there is very little resistance in the circuit, the needle swings full scale to the right indicating a short circuit; otherwise, it gives some indication of the amount of resistance between the points contacted. These points can be anywhere in the circuit, as

Fig. 1-20 Test points for continuity check

shown in Fig. 1-20. To start, the two leads may be connected to the prongs of the plug, points A and B in the
figure. If the needle swings toward the right, there is a
continuous circuit. However, this does not necessarily
mean that the circuit in the appliance is continuous, since
a short circuit in the line cord could cause the meter to
indicate continuity. By moving the switch to the *off*
position, the continuity should be broken. If it is, the line
cord is not shorting.

Checks can be made between points C and D to
determine if the switch is working. Here the indication
should be a short circuit when the switch is closed and an
open circuit when it is open. The resistance in the circuit
in Fig. 1-20 may be a heating element, a motor, or any
other electrical component. By checking between points
E and F you can determine if there is a break in the
wiring in this part of the circuit.

It is possible to use a test-light to make a continuity
test but it requires a lot of care since it involves the line
voltage as well. One lead from the test-light is plugged into
one slot of a receptacle and a long wire (2 or 3 feet) is
plugged into the other. The set-up is shown in Fig. 1-21.
This arrangement is sometimes called a *series test-light.*
Now if the two probes at the right are brought together,
the bulb will light, since the circuit will be complete.
Again, if you touch these probes to points C and D in

Basic Tests   **25**

TEST LIGHT

LONG WIRE

Fig. 1-21 Series test-light

Fig. 1-20, the bulb will light if the switch is closed, but will stay off if the switch is open. The series test-light can be used in the same manner and same places as the ohmmeter. Continuity is indicated by the lighting of the bulb.

Frequently when an appliance fails, the trouble will be obvious. For example, a switch fails to "click" when pushed to the *off* position, or a wire inside the appliance is obviously broken. You can make a quick check to determine whether the apparent trouble is in fact the actual trouble. For example, you suspect the switch. *With the appliance disconnected,* take a short clip lead and connect it across the switch. That is, clip one end to point C and the other to point D in Fig. 1-20. The clip lead will complete the circuit if the switch failed to do so. Now plug in the appliance. If it works, the switch must be at fault and should be replaced. In the case of the broken wire, use a clip lead to fasten the two ends together. Now when you plug in the appliance and turn it on, it should work if the only trouble was the broken wire.

Chapter 2

Safety

ELECTRICITY CAN BE LETHAL

The fact that electricity can kill you is something you must keep in mind whenever you work on any kind of electrical appliance. You might be lucky and escape injury if you carelessly touch a live wire. You might also avoid an accident if you drive an automobile the wrong way on a one-way street. Common sense dictates that both actions should be avoided. Many of the warnings in this chapter also appear in Chapter 1, but because of the danger involved, they bear repeating.

Always approach the repair of an electrical appliance with caution. This means giving the task your undivided attention and thinking constantly of the danger. A shock caused by carelessness is just as dangerous as one due to ignorance. You must also bear in mind that after the appliance is repaired, it must be safe to operate. That is, there must be no danger of shock to the person who uses the appliance in a normal manner. There is nothing you can do about someone who uses a metal knife to pry out a piece of bread stuck in a toaster while the toaster is on, but you *can* make sure that he won't get a shock if he touches the case of the toaster and the water faucet at the same time.

There are three simple rules of safety when working on appliances. If they are followed *at all times*, you will avoid danger to yourself and to the user.

1. *Do not touch any bare wire* which is connected to an electrical outlet.

2. *Never work on anything "live".*

3. *Always* check to make sure that there is no voltage between the case of the appliance and ground.

The first rule is fairly obvious and is usually carefully observed by beginners, but it is sometimes disregarded by sophisticated repairmen who should know better. It should never be disregarded. *Never* touch a bare wire which is connected to a source of voltage even if you *know* that it is safe. Consider the simulated appliance shown in Fig. 2-1.

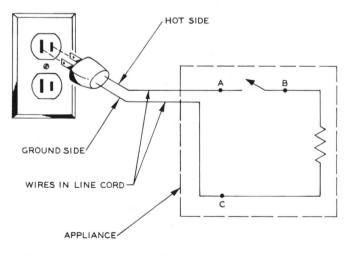

Fig. 2-1   Simulated appliance

Electricity Can Be Lethal

The line cord is plugged into a receptacle, but the switch in the appliance is in the *open* position. Point A to the left of the switch is now connected directly to the hot side of the line. Thus, point A has a potential of 115 volts from ground. Point A is a dangerous point to touch, especially if you may also be touching ground, (If you are standing on a concrete floor in leather-soled shoes, you are touching ground.) However, you know that point C is connected to the ground side of the line and should be safe to touch. Also, point B is connected to point C through the appliance and therefore should also be safe to touch. The trouble is that you know too much. It is true that you can safely touch points B and C in Fig. 2-1. This, however, is a bad habit to get into. In the first place, you might make a mistake as to which is the hot side of the line. Secondly, you might pull out the plug and then put it back upside down so that point A is now grounded, and points B and C are hot. Worse, someone else might do this while your attention is elsewhere. Therefore, never touch a bare wire unless it is disconnected from all voltage sources.

When you are making voltage checks with a meter or a test-light, it is necessary to hold the probes in contact with points of high voltage in the wall outlet or in the appliance being tested. Always hold the probes by the insulation. Again, never touch a bare wire or a bare probe.

The second rule, never work on a "live" appliance, refers to actual repairs. Of course, you must plug the appliance in and turn it on in order to make voltage checks on it. In this sense, you are working on a live appliance. However, once you have located the trouble, you must pull out the plug before making a single change or removing a single screw. Note that it is not enough to shut off the switch. If the appliance is plugged in, there are points of high voltage, (for example, point A in Fig. 2-1). Make it a habit to pull out the plug and *have it lying in plain view* while you are working on the appliance. In this way you will not get a shock because you *thought* you pulled the plug.

The third test, the ground test, checks the appliance for safety to the user. The purpose of this check is to make sure that the case of the appliance does not have a voltage on it. Before you do this, it is well to check your outlets to determine which side of the line is grounded. If the receptacle has been wired correctly, the longer slot is the grounded side of the line, as shown in Fig. 1-4. However, sometimes a homeowner decides to install a new receptacle himself and does so incorrectly. Then the short slot becomes the grounded side. Using a test-light or meter, put one probe in a slot in the receptacle and touch the other probe to a good ground connection, such as a water pipe. If the bulb lights or the meter reads about 115 volts, the probe in the receptacle is in the hot side of the line. Check the other side, anyway. The light will not go on if you have the wall probe in the slot connected to the grounded side of the line.

If a wall outlet is installed correctly, the mounting screw in the center is connected to ground through the shield that encloses the wires in the wall. Thus, the screw can be used as a ground connection when a water pipe is not available. This is illustrated in Fig. 2-2. One probe can be placed on the screw and the other in the hot slot, and the bulb will light. You must be sure that the screw is not covered with paint or other insulating material.

Fig. 2-2   Ground screw

To make the ground safety test, plug in your appliance and check to see if you have voltage between the case and a good ground. If you have ascertained that the screw on the wall outlet is a good ground, you can use that; otherwise, use a water pipe or faucet. The test is illustrated in Fig. 2-3. A toaster is shown in the figure, but the same test could be used for any appliance. One probe of the meter is held in contact with the case of the appliance and the other probe is connected to the good ground. In the figure, the screw in the center of the receptacle is used as a ground.

Electricity Can Be Lethal

Fig. 2-3   Ground test

If there is any voltage reading at all, even as little as one volt, there is a potential danger. A low voltage won't give you a shock, but it does indicate a faulty connection or weak insulation, and it can only get worse. The trouble should be corrected before it becomes dangerous. Note that a normal test-light may not be sensitive enough to detect a low voltage, but a neon tester will. If you do not have a meter, a neon tester is a necessary tool along with a test-light.

In making the ground test, make sure that you pull out the plug and invert it so that the grounded wire in the line cord becomes the hot wire, and vice versa. Make the test for both positions of the plug with the appliance switched on and with it off. There should be no voltage between case and ground in any of these tests. If there is a voltage in even one position of the plug and switch, the appliance is not ready for use until the cause of the trouble is eliminated.

One way of making certain that the case of an appliance will never be hot is to permanently fasten the case directly to ground. This should be done on all large appliances that are never moved, such as an oven, clotheswasher, or clothes dryer. If you have a wall receptacle with a ground terminal as shown in Fig. 1-5, then it is a simple matter to run a three-wire line cord to the appliance. The third wire is fastened to the case at the appliance end and to the round prong on the plug. Even without this type of grounding outlet, you can still ground the appliance by running a wire from the case to a ground connection. Use a piece of wire which is rated to carry the current drawn by the appliance. Loosen a screw on the case (preferably in back and out of the way), wrap an end of the wire around the screw, and tighten the screw again. Connect the free end of the wire to a grounding clamp which is attached to a water pipe. Grounding clamps are available in hardware or electrical supply stores. The case of the appliance will then be firmly connected to ground. If subsequently something shorts inside the appliance putting a voltage on the case, it may cause a fuse to blow, but it will not endanger the life of someone touching the case.

Repairing electrical appliances must not be done haphazardly. Neatness is important, not only because it looks good, but because it is safer. For example, a clean connection has no stray wire sticking out which may contact the case of the appliance or touch another wire causing a short circuit. Connections should be "solid" so that the appliance will stand the abuse that careless users too frequently give them. Although everyone knows that a line cord should not be yanked out of an outlet, people do so frequently. Consequently, wiring must be done in such a way that this type of abuse will put no strain on the electrical connection. It is also obvious that wires must not interfere with moving parts in appliances.

Underwriters' Laboratories is the name of an association which formulates good manufacturing practices for all electrical products, including materials and components, as well as appliances. This organization also tests manufactured appliances to ensure that they meet the

Fig. 2-4   Underwriter's
Laboratories Seal of Approval

specified standards. Manufacturers are permitted to indicate that their appliances conform to the standards by placing an "Underwriters' Seal" on their appliances that do conform. The most common form of this seal is shown in . Fig. 2-4. All appliances in the home should bear this seal. If you look at Fig. 1-7 and 1-8, you will notice that every nameplate illustrated bears this seal.

Chapter 3

Line Cords

The most common source of trouble in an electrical appliance is the line cord and its connectors. Although the repair or replacement of line cords and the installation of new connectors are relatively simple tasks, more often than not they are repaired improperly, leading to further trouble. It is important to know something about wires and connectors to be sure that you select the proper material for the job.

CORDS

Since line cords must be flexible to go around objects or to permit movement (with vacuum cleaners and electric irons, for example), they are always made of stranded wire. If a single solid wire were used for each conductor, it would have to be large enough to carry the current required by the appliance. For most appliances, the wire size required could be achieved only in a stiff wire. Thinner wire is flexible, but will not carry sufficient current without overheating. However, if several thin wires (or strands) are grouped together in a single conductor, they share the current equally, and thus can carry heavier currents without overheating. All line cords use stranded conductors. Cross-sections of individual conductors are shown in Fig. 3-1. Each conductor consists of a bundle of strands of thin wire. This bundle is tightly enclosed with a wrapping of fiber, called a *lay,* and then covered with rubber insulation. There are several different types of line cords, but all those used with home appliances have the conductors wrapped in fiber and rubber, as shown in Fig. 3-1. The more popular line cords have conductors with seven or ten strands, but for heavy duty installations wires with as many as 65 strands in each conductor are used.

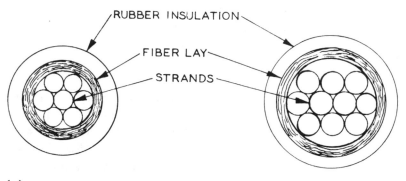

(a) 7-STRAND   (b) 10-STRAND

Fig. 3-1   Stranded conductor with wrapping

When you strip insulation from a conductor in a line cord, it is important that you do not cut into any of the strands. If you break a few strands, the appliance will operate properly, but each of the remaining strands will carry more current than it should. For example, suppose a toaster which draws 10 amperes has a line cord with 10-strand conductors, as shown in Fig. 3-1 (b). Each strand then carries only one ampere, and these strands

Fig. 3-2 Stripping insulator

can be relatively thin wires. However, if six strands break because of too much flexing or poor installation, the remaining four will have to carry the 10-ampere load, and each will carry 2.5 amperes. This will cause the line cord to get hot in the vicinity of the break. You can use a jack-knife to strip insulation, but you should not use too much force. Cut into the insulation *lightly* by rotating the

Fig. 3-3 Zipcord

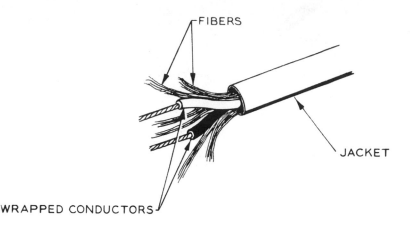

FIBERS

JACKET

WRAPPED CONDUCTORS

Fig. 3-4   SV cord

knife around the wire about an inch from the end, as shown in Fig. 3-2. When the insulation is cut all the way around, the severed portion can be pulled off like a sleeve. The fiber lay between the strands and the rubber insulation protects the strands from the knife, and there should be little risk of cutting the wire strands as long as the job is done carefully. After pulling off the sleeve, the fiber lay can be unwrapped, as shown on the conductor at the left in Fig. 3-2, and can be cut off with a pair of scissors or a knife.

For most low-power appliances, such as clocks, juicers, and lamps, which draw less than two or three amperes, the most common type of line cord is the popular *zipcord* shown in Fig. 3-3. This consists of two conductors, like those in Fig. 3-1, joined together by a thin rubber bond. It is called zipcord, because the two conductors are easily separated by tearing the thin rubber membrane joining them.

For heavier work, such as mixers and vacuum cleaners, *jacketed cable* is used. The wrapped conductors are separated by additional fibers which increase the insulation between them, both electrical and thermal. The two conductors and extra fibers are encased in a rubber or plastic jacket. This type of line cord is called SV cord, and its construction is shown in Fig. 3-4. In general, jacketed cable will take more abuse than zipcord.

For appliances which draw more than five amperes, special *heater cord* is preferred. This is used on toasters, waffle irons, irons, and other appliances furnishing heat rather than motion. In heater cord, each wrapped conductor shown in Fig. 3-1 is further wrapped with asbestos fibers. The two conductors are then brought together and a layer of asbestos is wrapped around both. Finally the package is encased in a braided jacket of cotton or nylon. Heater cord is illustrated in Fig. 3-5.

The current-carrying capacity of line cord depends on the wire diameter and not on the insulation. Thus, a zipcord line with #10 wire can carry 25 amperes, whereas an asbestos heater cord with #18 wire may be limited to 5 or 6 amperes. In general, heater cords are used for appliances that draw more current; jacketed cable is used where the cord itself is subject to abuse; and zipcord use is by far the most common because of its simplicity. Zipcord

(a) WITH JACKET
CUT BACK

(b) WITH ASBESTOS
UNWRAPPED

Fig. 3-5   Heater cord

Cords

(a) MOLDED PLASTIC

(b) SHELL TYPE WITH SCREWS

(c) GRIPPER

(d) WALL HUGGER

Fig. 3-6   Plugs

is used wherever there are no special demands on the line cord.

CONNECTORS

Most of the early appliances did not have line cords connected to them, but instead a *cord set* was supplied with each appliance as a separate unit. A cord set consists simply of a line cord with a connector at each end, one a wall plug to connect to the wall outlet, and the other an appliance plug to connect to the appliance. Some appliances, such as irons and vacuum cleaners, are moved frequently in use, and this motion causes excessive bending and strain at the appliance end of the cord, which often causes the wire to break inside the connector or to become disconnected. To eliminate this source of failure, manufacturers now tend to have the appliance cord fastened to the appliance internally, although a few appliances still come with separate cord sets. Another advantage of an integral line cord is that a user cannot connect the wrong cord set to an appliance.

Connectors

Wall plugs come in a large assortment of shapes, materials, and colors. A few types are shown in Fig. 3-6. The molded plastic plug in (a) is usually furnished with new appliances. The connections between wires and prongs are tightly sealed, and with proper use this type of plug should outlast the appliance.

The shell-type plug shown in (b) is a replacement plug which can carry high currents. Each individual wire in the cord is attached to a screw in the plug (as described later in this chapter). This type of plug is shown in two shapes, rectangular and round. The round shape is more common. The shell may be either plastic or rubber.

The gripper-type plug shown in (c) is the simplest to install, but is suitable for use only on zipcord for low-current applications. To attach this type of plug, you simply cut the zipcord squarely, and without stripping off any insulation insert it in the hole in the side of the plug as far as it will go. Then push down the lever on the back of the plug, which causes small teeth on the connectors to bite through the insulation and make contact with the wires. The *wall-hugger* plug shown in (d) is the same kind of connector, but has the prongs protruding from the largest face, so that when it is inserted into a wall receptacle, the plug extends less than half an inch from the wall and causes a minimum of interference with furniture next to the wall. Wall-hugger plugs are also available with screw connections for handling higher currents.

An appliance plug is shown in Fig. 3-7. This plug comes in two halves which are either held together with screws and nuts or rivets. Inside are two heavy-duty contacts which slide over projecting pins in the appliance.

Fig. 3-7   Appliance plug

These contacts have screws to which the wires in the cord are connected. The spring at the back of this plug prevents sharp bends in the cord which may damage the insulation or the wires.

FAULTS AND REPAIRS

Many old line cords on appliances were covered with re- claimed rubber which does not stand up well over a long period of time. If these cords are exposed to sunlight for long periods or stretched behind radiators, the rubber becomes dry and brittle, and sometimes pieces break off in chunks leaving exposed wire. Needless to say this dan- gerous condition must be corrected as soon as it is detected. If a visual inspection of any cord shows cracks or dryness in the insulation, you should flex the cord sharply to see if the insulation can stand it. If not, replace the cord.

Modern line cords should outlive the appliances they are attached to. However, modern users of line cords tend to abuse the cords and cause problems. The most common fault is pulling a plug from the wall outlet by yanking on the cord. There are two dangers here. You may pull the wire from its screw contact, if it was improperly connected, or, secondly, you might break some or all of the strands in a wire. Another problem arises when something is pushed against a plug in a wall receptacle so that the cord entering the plug is bent sharply. If this happens too often, the wire may break from too frequent bending. Other obvious problems occur when someone accidentally steps on a plug, shattering it, or accidentally cuts a line cord.

When an appliance fails to work, your first check should be to test the wall outlet, as illustrated in Fig. 1-16 on page 21 and make sure there is voltage at the outlet. If there is not, check your fuse box. If there is voltage at the receptacle, but no voltage across the line at the appli- ance end of the cord, then the fault is in the cord or one of its connectors. If a cord is broken, it can be replaced. However, if the break is near the wall plug, you can simply cut off the broken end of the cord, and attach the plug to the shortened cord remaining.

A more difficult problem is to find the source of trouble when an appliance sometimes works and sometimes doesn't. This is called an *intermittent fault,* and it usually can be traced to a break in the line cord. Whenever the two severed ends of the conductor are in contact because of the way the cord is bent, continuity will be established and the appliance will work. If the fault is in the line cord, it can be found by switching the appliance to the *on* position and flexing the line cord slowly beginning at the plug end and progressing toward the appliance. If the intermittent fault is caused by a break in the cord, the appliance will operate at some positions of the cord as it is bent, but not at all positions. From this you can determine where the break occurs. If near the plug, proceed as before; if not, replace the cord.

When some of the strands in a cord are broken, but at least one remains intact, current will still flow, but the current through the remaining strands may exceed their

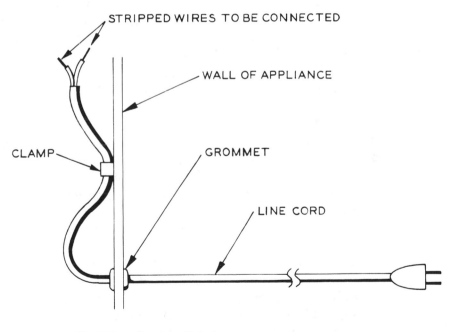

Fig. 3-8 Strain-relief clamp

rated current-carrying capacity with the result that the wire gets very hot at the point of the break. This heat is caused by current flowing through the increased resistance of the smaller wire and is evidence also of a drop in voltage. The result is lower voltage at the appliance and consequently less heat produced there. Although this would mean an increase in time to toast a slice of bread, or brew coffee, for example, it is unlikely that the average householder would notice this. However, the hot spot on the cord should be corrected as soon as it is detected by cutting off the injured section if it is near the plug or by replacing the whole cord.

When a cord is shortened or replaced, it is necessary to attach a plug to it. It is important to connect a cord to a plug and to an appliance in such a way that the connection will not be subject to strain when the cord is pulled. At the appliance end, the cord is usually clamped a few inches away from the connection, as shown in Fig. 3-8. When attaching a line cord to a plug, you can also remove strain from the connections by tying a knot in the line cord after passing it through the connector. The typical "Underwriters' knot" is shown in Fig. 3-9. First you slip the plug over the line cord as shown in (a). The first loop of the knot is shown in (b), with the rest of the knot indicated by a dotted line. The knot is slid down to the end of the insulation. Any kind of knot will do as long as it prevents the cord from being pulled out of the plug. The two wires are now stripped, passed around the blades of the plug, and the stripped portion tightened under the mounting screws, as shown in (c). The wire should not be crossed over itself under a screw, as tightening may cause it to break.

If the appliance draws less than two amperes and has a zipcord line cord, you can use the gripper plug shown in Fig. 3-6 (c). Here there is no strain-relieving knot, no stripping of wire, and no contact screws. Although the connection can be damaged by jerking the line cord, with reasonable care it can give adequate service indefinitely. There are many varieties of gripper plugs on the market, and all come with simple instructions for installing. *Warn-*

(a) PLUG SLIPPED OVER CORD

(c) CONNECTED

(b) KNOT

Fig. 3-9 Strain relieving knot

ing: do not use gripper plugs on anything but zipcord and in low-current applications.

Heater cord needs some special treatment. After the heater cord is stripped, as in Fig. 3-5(b), there is no longer asbestos between the two conductors in the area outside of the jacket. To remedy this, after stripping and before proceeding further the asbestos fibers shown hanging over in the figure should be brought up around the rubber insulation of the two leads and fastened there by tying with a piece of thread. Then this cord can be fastened to appliance and plug in the usual manner.

Whenever you have occasion to disconnect a cord at the appliance, note carefully how the cord is routed and where it is connected, so that you can fasten a new cord in the same manner. It is useful to draw a sketch. This is especially important when a part with many connection wires is removed. A sketch of all connections, showing colors of wires, is almost a necessity.

Faults and Repairs

When an appliance plug is to be connected to a cord set, the plug is taken apart by unscrewing the two screws holding it together. If the two halves are riveted together, you might do well to cut off the plug and buy a new one that uses screws. However, it is possible to drill out the rivets and replace them with screws when putting the plug back together again. In either case, inside the plug are two heavy duty connectors. The two wires of the line cord must be stripped and connected to screws on these connectors. First, you must slip the strain-relieving spring onto the cord as it will not go over the connectors. After the wires are connected, the connectors and an end of the spring are placed in appropriate grooves that are cut out to receive them, and the two halves are screwed together. It is *not* necessary to tie a knot in the cord since the interior of the appliance plug is designed to give strain relief without it.

Chapter 4

Heating Elements and Thermostats

The simplest electrical appliances and thus the easiest to service are those that produce heat. All contain some sort of heating element which gets hot when an electrical current passes through it. If an appliance is *automatic,* that is, if it maintains a certain temperature without manual adjustment, then it usually contains a thermostat as well.

HEATING ELEMENTS

Heat is produced when an electric current is passed through a resistance. The power dissipated in the resistance, as discussed in Chapter 1, is the product of voltage and current. That is,

Watts = Volts x Amperes.

The resistance of a wire is expressed *ohms,* which is the ratio of volts to amperes. That is,

Ohms = Volts/Amperes

Amperes = Volts/Ohms

These formulas are usually expressed as follows:

$$P = E \times I$$

$$R = E/I$$

$$I = E/R$$

where P is the power in watts
E is the electromotive force (EMF) in volts
I is the intensity of the current in amperes
and   R is the resistance in ohms.
For example, assume the line voltage is 115 volts and the current through a toaster is 10 amperes. Then,

Power = 115 x 10 = 1150 watts

Resistance = 115/10 = 11.5 ohms

It should be noted that when a toaster is turned on, current flows through the line cord and the heating element in series, as depicted in Fig. 4-1. The line cord has a very

Fig. 4-1   Series resistance

small resistance, much less than 0.05 ohm. The heating element for the 1150-watt toaster has a resistance of 11.5 ohms as shown in the example above. The two resistances are said to be in *series* since the same current flows through both. The 115 volts divides across the resistances in proportion to the values of resistances. That is, since 11.5 is more than 200 times 0.05, we would expect the voltage across the heating element to be about 200 times the voltage across the line cord. Thus, we might expect to measure about half a volt across the line cord and about 114.5 volts across the heating element. The figure of 115 volts for the line voltage is only an approximation, since the line usually varies from 110 to 120 volts, depending on the total load at the power station. Since the line cord resistance is so small compared to that of the heating element, we can neglect it and assume that the 115-volt line potential is entirely across the heating element.

Note, however, that if the line cord has some strands broken so that its resistance increases to only one ohm, the new division of voltage could be detrimental. Now the one-ohm line cord is in series with the 11.5-ohm heating element, and the total resistance across the line is 12.5 ohms. The current is then

$$115 \text{ volts} / 12.5 \text{ ohms} = 9.2 \text{ amperes}$$

The voltage divides according to the resistance. That is, the voltage across the line cord is 1/12.5 of the total voltage, and the voltage across the heating element is 11.5/12.5 of the total voltage. This means that the voltage across the heating elements is only 105.8 volts. The power dissipated in the heating element is then

$$105.8 \text{ volts} \times 9.2 \text{ amperes} = 973.36 \text{ watts}$$

This is about 85 percent of the rated power of 1150 watts and means that the toaster would produce only 85 percent of the heat it should. This is another reason why it is very important to make sure line cords are in good condition.

The resistance of the heating element determines the power that will be used. Thus, if you want a 1000-watt element, you know that the current through it should be 1000/115 or 8.70 amperes (current = watts/volts). Since resistance is volts divided by amperes, you know that the resistance should be 115/8.70 or 13.2 ohms. Therefore, you need a 13.2-ohm heating element to dissipate 1000 watts. Values of current and resistance for several power ratings are given in Table 4-1.

Since ordinary copper wire has very low resistance, it is used in line cords and other parts of the circuit where heat is *not* wanted. Special materials with much higher values of resistance are used where it is necessary to produce heat or light. Most of these materials produce both heat and light when power is dissipated in them, but they are chosen for the particular application on the basis of which they produce better. Thus, for example, tungsten is used for light bulb filaments because it produces light well, although it also gets hot. Similarly, a heater element in a toaster glows when it gets hot, but heat is its most important output. From Table 4-1, a 500-watt unit should have a resistance of 26.4 ohms. This is true whether it is a 500-watt bulb giving off light or a 500-watt heater producing heat.

The most common material used in heating elements is *nichrome,* an alloy of nickel and chromium. For the same size wire, nichrome has a resistance more than 1000 times that of copper. In toasters, flat nichrome wire or ribbon is used as the heating element since it can be maintained at a fixed distance from the bread to be toasted. Nichrome ribbon has a resistance of about 1/20 ohm per inch, which means a length 20 inches long has only one ohm resistance. From Table 4-1, the resistance for a 1000-watt toaster must be 13.2 ohms, and since each ohm requires 20 inches of ribbon, the total length of the heating elements must be 264 inches or 22 feet. There are usually four elements in a toaster, one on each side of each piece of bread, so that each of the elements must be one-fourth of 22 feet or 5.5 feet. This length is accommodated in the toaster by winding the ribbon back and forth on mica supports. Figure 4-2 is a cutaway sketch of a toaster, showing the nichrome ribbon heating elements.

TABLE 4-1 CURRENT AND RESISTANCE AS A FUNCTION OF POWER

(115-volt line)

Power (watts)	Current (amps)	Resistance (ohms)
500	4.35	26.4
750	6.52	17.6
1000	8.70	13.2
1250	10.87	10.9
1500	13.04	8.82

In order to reduce the length of the heating element, the nichrome wire is frequently coiled on a one-quarter inch diameter. This is the form of heating element used

Fig. 4-2   Toaster heating element

Fig. 4-3 Coiled nichrome heater

in hotplates, as shown in Fig. 4-3, as well as inside waffle irons, grills, and similar appliances. Note that the heating element is one continuous coil from point A to point B. Coiled nichrome wire has a resistance of approximately 0.4 ohm per inch of coil. Using this value, a 1000-watt appliance would need a heating element 33 inches long as compared to the 264 inches of nichrome ribbon. Lengths of heating elements for both nichrome ribbon and coiled nichrome wire are given in Table 4-2. Note that as the length of the element is decreased, the resistance also is decreased, and the power is increased. This means that if a heating element broke and you tried to make a temporary fix by discarding the shorter piece and stretching the longer piece to join it to its terminal, you would be using a lower resistance and would thus have more current and more power. You would actually get more heat, but should be careful not to exceed the current rating or power rating of other parts of the circuit.

Both nichrome ribbon and coiled wire are used in heating elements which are enclosed so that the user will not be apt to touch the elements. Note that there is a potential danger of shock from touching a heating element and ground at the same time whenever the element is connected to the line, even if it is not turned on. This is explained in Fig. 4-4. The switch in the circuit is usually in one side of the line only. (In a toaster it is actuated by pushing down the plunger.) If the line cord is plugged into a wall outlet, so that the switch is in the grounded side, then the heating element is connected directly to the hot side of the line. If you touch this and are grounded at the same time, you will get an electric shock just as if you touched both sides of the line simultaneously. If you must use a knife to pry out a piece of bread stuck in a toaster, make sure you pull the wall plug first.

TABLE 4-2   LENGTH OF NICHROME HEATING ELEMENTS

Power (watts)	Resistance (ohms)	Nichrome ribbon 0.05 ohm per inch (inches)	Coiled nichrome wire 0.4 ohm per inch (inches)
500	26.4	528	66
750	17.6	352	44
1000	13.2	264	33
1250	10.9	218	27
1500	8.82	176	22

Heating elements that must be exposed, as for example on the top of a range, are usually made of coiled nichrome wire completely enclosed in a metal sheath, from which the nichrome wire is insulated electrically. The trade name for one such type of heating element is Calrod, and this name is usually applied indiscriminately to all sheathed heating elements. Calrod heaters as used on top of a kitchen range, are shown in Fig. 4-5. The metal sheath also protects the nichrome wire from oxidizing.

Heating elements are usually connected to their terminals with bolts. Solder could not be used, since it would

HOT SIDE OF LINE

HEATING ELEMENT

GROUNDED SIDE
OF LINE

SWITCH

Fig. 4-4   Potential shock hazard

melt from the intense heat. If a heating element breaks, you should replace it with one of the same electrical rating and physical size. Sometimes a temporary repair can be made by fastening the two broken ends together with a bolt and nut as shown in Fig. 4-6. Each of the two wires is wound around a small bolt, and flat washers are used to maintain even pressure. The bolt must be insulated from the rest of the circuit and from anything that could be in contact with the user. Note that this is strictly a *temporary* solution, and the heating element should be replaced as soon as possible.

THERMOSTATS

A *thermostat* is a switch which is controlled by heat. The most common type used in electric appliances is a *bimetallic*

Fig. 4-5   Calrod heating elements

Fig. 4-6 Quick fix of heating element

thermostat which consists of two strips of different metals firmly fastened together, as shown in Fig. 4-7. One of the metals has a large rate of thermal expansion, while the other has low thermal expansion. When cold, the combined strip is straight, as shown in (a). When the unit is heated, the two metals expand at different rates, and the alloy with a high coefficient of thermal expansion becomes longer than the other. This results in the unit bending, as shown in (b). When the thermostat cools it returns to the configuration in (a).

A thermostat can be used to make a contact or to break one. In Fig. 4-7, two terminals, A and B, are shown, one on each side of the thermostat. Two contacts are also shown on the thermostat itself, C and D. When the thermostat is cold, as in (a), contact D touches terminal B, and connects B to the circuit (represented by the wire screwed to the left end of the thermostat). When heat is applied, the thermostat bends, breaking this connection, and now contact C connects terminal A to the circuit, as shown in (b). In practice, only one contact and one terminal are usually used. If the thermostat is used to break a circuit when it gets hot, then only contact D and terminal B are used. When cold, the thermostat connects these two together, and when hot, it breaks this connection. On the other hand, if the thermostat is needed to turn on a circuit when it gets hot, then only A and C are used.

Thermostats

METAL WITH LOW RATE OF EXPANSION

METAL WITH HIGH RATE OF EXPANSION

(a) COLD

(b) HEATED

Fig. 4-7   Thermostat

The temperature at which the thermostat makes or breaks a connection can be controlled very accurately by the spacing between the contact and the terminal. In some appliances, such as waffle irons and coffee percolators, it is necessary to bring the heat up to a certain specified temperature and to maintain it there. At the start, the thermostat closes the circuit since it is designed to make a connection when cold. As the temperature increases and the thermostat bends, it finally reaches a point where it breaks the circuit by separating contact D from terminal B. Since there is some springiness in the metals, contact B can be pushed up so that the break occurs at the correct temperature. In many appliances, this temperature may be varied by turning a knob or moving a lever which moves contact B up or down. On a waffle iron or toaster, this knob may have a pointer which is set on a variable scale from "light" to "dark". In any case, as soon as the contact is broken, the heating element in the appliance no longer operates, and the temperature starts to decrease. Now the thermostat bends back and closes the

Thermostats

connection again, and the whole cycle is repeated. By continually making and breaking the contact in this way, the thermostat holds the temperature in the appliance within narrow limits. By suitable choice of alloys it is possible to design a thermostat to maintain a constant temperature in an appliance within one degree, although in practice five or ten degrees variation is acceptable.

Thermostats are used in most appliances using heat, such as electric blankets, toasters, waffle irons, irons, and others. Wherever a variable temperature is desired, this is set by controlling the spacing of the contacts. The operating temperature can be changed over a wide range in this manner.

Chapter **5**

Small Heating Appliances

The main purpose of the electrical portion of many home appliances is simply to produce heat. The heat may be used for cooking or warming food, as in a toaster or electric skillet, or for warming the person or clothes, as in an electric blanket or electric iron. In any case, the electrical circuit of the appliance is very simple, consisting of a line cord, a heating element, a thermostat, and some-times a switch. The operation of many of these appliances and repair techniques used when they fail are presented in this chapter.

Before attempting to repair any appliance you should be familiar with the basic principles of appliance repair presented on page 13 and with the two basic tests explained on page 19. It may be worthwhile rereading these sections, since in this chapter and those that follow it will be assumed that you are familiar with these principles and tests.

HEATING PADS

Heating pads are used to apply local heat to various parts of the body, usually to relieve pain. The pads may be as small as a foot square and range to two or three times this size. They must be flexible in order to conform to the shape of the joint or other part of the body which needs warmth. The heating pad is usually encased in a washable cover. The pad itself contains two heating elements, one typically rated at 20 watts and the other at 40 watts. These are flexible, thin wire elements which are sewn right into the material forming the pad. The two elements wind back and forth in the pad in parallel. At one end they are joined together and a *common* lead from the junction runs to the control box. The free ends of the elements are each connected directly to a separate terminal on the control box.

A typical control box for a heating pad is shown in Fig. 5-1. Note that a three-wire line runs from the control box to the pad, and a two-wire line is used as the line cord. These lines are usually made of zipcord. One lead of the two-wire line cord goes right through the control box and is the common lead for the two heating elements. The other lead runs to the center contact of a rotary switch. The horizontal bar shown at the center of the control box in the figure is used to rotate this switch. With the bar as shown, the switch is open, and thus the heating pad is off. When the switch is rotated counterclockwise so that one end of the bar points to L, an arm of the switch makes contact with the wire leading to the *low* heating element. The circuit for this element is closed, and current flows through it, producing heat. Rotating the bar so that it points to M moves the contact from the *low* element to the *medium* element, producing approximately twice as much heat. In the *high* position, when the bar points to H, both elements are connected simultan-

Fig. 5-1 Heating-pad control box

eously. The dots shown on the control box are embossed on the case so that they may be felt in the dark. This makes it possible to set the control "by feel" without looking at it. When one end of the bar points to L, the other points to a single dot. Two dots correspond to M, and three to H.

The control box is held together by one or two screws which can be removed to gain access to the ends of the two-wire and three-wire zipcords. Voltage checks at the junctions in the box can be used to determine if the line cord is all right and if the switch is operating properly.

Alternatively, the switch can be tested by continuity checks. A continuity check is also used to determine whether there is a break in a heating element.

If the heating pad will not get warm in any position of the switch, the trouble is probably *not* in the heating elements, since it would be surprising for both to fail at the same time. Check the switch and the line cord. Also check the connections of the wires to the switch. If the wire in the line cord running to the center contact becomes disconnected, for example, the circuits will remain open with the switch in any position. If all connections are satisfactory, make a voltage check in the control box to make sure the line cord is satisfactory. No voltage means a faulty line cord or a dead outlet.

If the fault is in the line cord or is due to a break in the three-wire zipcord, it can be repaired by splicing in a small piece of wire joining the two ends of the break. This is permissible, since this appliance draws low current. If possible, solder the extra wire in place at both ends, but if no soldering iron is available, a good twisted splice will

CONTROL
BOX

COMMON LEAD

HEATING ELEMENTS

Fig. 5-2   Electric blanket

do. In any case, cover the joint well with insulating tape to prevent danger of shock from bare wires.

If the fault is due to a break in a heating element inside the pad, there is little you can do to fix it. The heating pad may still be serviceable using only one heating element, so it should not be discarded, but you should not try to repair it.

In some heating pads, a small thermostat is placed inside the pad itself, in series with the heating elements. If this thermostat fails in the closed position, it should be ignored, and the pad can be used in the same manner as those without thermostats. If the thermostat fails in the open position, you may be able to remove it or replace it. It is necessary to rip open a seam of the heating pad and find the thermostat. Usually, the bad thermostat is removed completely, and the two wires which were connected to it are simply tied together and *carefully insulated* before sewing up the seam again.

ELECTRIC BLANKETS

An electric blanket is very similar to a heating pad in that it contains flexible heating elements sewn into the material. There are usually two heating elements, one for each half of the blanket. As with the heating pad, a three-wire line runs from the control box to the blanket. This line usually has a connecting plug and receptacle on it so that the control box can be separated from the blanket. A typical circuit is shown in Fig. 5-2.

In a single-control electric blanket, both heating elements are connected simultaneously by one control and the temperature is regulated by one thermostat which is inside the control box. In a dual-control blanket, there are two separate controls on the control box, one for each heating element. In some models, two separate control boxes, each with its own thermostat, are used with a dual-control blanket.

The temperature is controlled automatically by the thermostat in the control box. This thermostat has a mag-

Fig. 5-3   Thermostat

netic contact, as shown in Fig. 5-3. The lower contact is mounted on the base of the control box and is surrounded by an iron washer. The upper contact is in the center of a cup-shaped magnet, and both are attached to the bimetallic arm of the thermostat. When the temperature *in the room* falls below a predetermined value, the thermostat arm straightens, allowing the magnet to come close to the iron washer. When it is close enough, magnetic attraction draws it the rest of the way, joining the contacts. Without the magnet, the contact points would bounce and not make a stable contact. When the temperature increases, the arms tend to separate but are held together by the magnets until the force of the bending thermostat arm overcomes the magnetic attraction. As a result electrical contact is made firmly and broken cleanly.

The original spacing between the two arms of the thermostat is adjusted by a knob on the cover of the control box. This sets the operating temperature. In addition most boxes contain an on-off switch and a pilot light, which is usually a neon bulb.

Most troubles with electric blankets occur in the control box or line cord. The knob can be pulled off and the box opened. In some cases, the knob is held in place by a set-screw which must be loosened. Once you have the box open, you should make a visual check for loose connections or broken wires. Then make voltage tests to make sure the voltage is getting to the box, and with the plug pulled out make a continuity test to make sure the heating elements are intact. Push down the magnet to check if it is drawn to the iron washer. After many contacts, a magnet can lose its magnetism. If so, replace it with a new magnet. Or if necessary replace the whole thermostat.

If a heating element opens, it cannot be fixed. Half of the blanket will still work if the unit is a two-element electric blanket. However, even if no heat is produced don't discard the blanket, since it can still be used as a regular blanket without electricity.

HOTPLATES

The simplest electrical appliance is a hotplate, consisting of a line cord and a heating element. The line cord may be permanently connected to the hotplate or may be a separate cord set. The hot plate is used to heat water in a pan or to do simple cooking where no stove is available. The heating element is usually made of coiled nichrome wire, as shown in Fig. 4-3 on page 52, but Calrod elements are sometimes used.

If a simple hotplate fails to operate, the trouble is either in the line cord or the heating element. This can be pinned down quickly by making one voltage check, as illustrated in Fig. 5-4. The line cord is plugged into a live socket and the voltage is checked at points A and B in the figure. These are the connections between the line cord and the heating element. The probes of your test-light or meter are held in contact with points A and B. If a voltage is present, the lamp lights or the meter indicates the voltage, and therefore the line cord is not at fault. The trouble must be in the heater. With no voltage, the trouble is in the line cord. A corroborating test if the trouble seems to be in the heating element is a continuity check. With the plug pulled out of the wall outlet, check the continuity between points A and B in Fig. 5-4. If the heater is open, there will be no continuity. In the case of an open nichrome coil

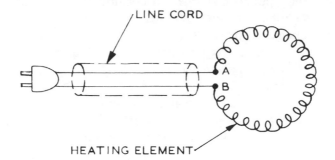

LINE CORD

A
B

HEATING ELEMENT

Fig. 5-4 Voltage check-points

heating element, you should be able to spot the broken wire.

Once the source of trouble is known, fixing the appliance is simple. If the cord is at fault, correct it as described in Chapter 3. If the heating element is bad, replace it with an identical element if possible.

To operate the simple hotplate, you plug it in to heat, pull out the plug to shut it off. A slightly more sophisticated model includes a switch in addition to the heating element and line cord. This unit can be left plugged in all the time and is turned on by flipping the switch. The switch is an additional source of trouble and should be checked if the hotplate fails to operate. With the plug pulled out of the receptacle check continuity across the switch both in the off and on positions. The switch should be open in the off position and closed (continuous) in the on position. If the switch seems at fault, clip a short clip lead across it, thus removing it from the circuit. Now when you put the plug back in the wall outlet, the hotplate should begin to heat. This is another confirming test. Do not leave the clip lead in place. Replace the switch.

Hotplates are also made with two or more heating elements and a separate switch for each. In effect, such a multiple-burner hotplate is really several separate appliances in one package. If all burners fail to work, the trouble is in the line cord or line voltage (check the fuse). If only one fails, the trouble is in that burner or its associated switch, but the line cord is not at fault.

Hotplates

In summary, the possible faults in a hotplate lie in the line cord, switch, or heating element, or their interconnections. If the hotplate blows a fuse when it is plugged in, the trouble is a short circuit, probably in the line cord. If the hotplate fails to heat, the trouble is an open circuit or a faulty switch. In any case, voltage checks and continuity checks will enable you to locate the trouble.

POPCORN POPPERS

A popcorn popper consists of a covered pan, which may be made of metal or glass, supported a suitable distance above a hotplate. Oil and corn are placed in the pan. The heat raises the temperature of the oil and causes the corn to pop. If the pan were too close to the source of heat, the corn might burn before the oil reached the proper popping temperature. The pan is usually supported by a flange around the hotplate, as shown in Fig. 5-5.

Fig. 5-5   Popcorn popper

SPRING

COIL

THERMOSTAT

Fig. 5-6   Waffle iron

The electrical portion of the popper is nothing more than a simple hotplate. The possible troubles and their corrective measures are discussed in the preceding section. In some models, referred to as *automatic* poppers, a thermostat is added to shut off the popper when the temperature gets too high, which usually happens after the corn is popped. The thermostat should be treated simply as a switch. It should be closed when it is cold so that the heating element begins to heat immediately when the popper is plugged into an outlet.

One possible source of trouble is a thermostat which opens at too low a temperature so that the oil never gets hot enough to pop the corn. If the oil in an automatic popper gets warm, but nothing else happens, the trouble is probably the thermostat opening too soon. Try bypassing the thermostat. If this corrects the trouble, you can leave it that way, since it will operate quite well as a corn

Popcorn Poppers

popper, although it will no longer be automatic. When you get around to it, you can replace the thermostat and make the popper automatic again. If bypassing the thermostat does not cure the trouble, the fault is probably a broken wire in the heating element. When cold the two ends of wire touch, completing the circuit, and the element gets hot. However, as the heater gets hot, it expands, opening the break. In effect, the broken heater is acting like a thermostat. If this is happening, you should be able to spot the break when the heater is getting hot. To fix it, replace the heater element.

A simple hotplate, with or without a thermostat, is also the only electrical part in a few other electrical appliances, notably old-fashioned vacuum coffee pots and deep-fat fryers. The possible faults and corrective measures for these appliances are essentially the same as for the simple hotplate described in this and the preceding sections. For the most part, manufacturers have done away with the separate hotplate, and modern appliances have the heating elements built into them.

WAFFLE IRONS

The electrical portion of a waffle iron consists of a line cord, switch, thermostat, two heating elements, and pilot light. The two heating elements, one in each half of the iron, are wired in parallel so that they are turned on and off together. The pilot light may be a small bulb, a neon light, or in some inexpensive models, simply a small window on the outside of the case. When a window is used, the glow from a heating element is visible and serves as a pilot light.

The heating elements in a waffle iron are usually coiled nichrome wire supported on porcelain insulators. Each element is mounted in a removable metal pan. The two elements are connected by wires passing through a protective spring on the hinge side of the iron. A waffle iron with bottom grill removed is shown in Fig. 5-6. Note the coil spring joining the two halves of the iron. Without this spring, the wires joining the two heating elements would be subject to too much abuse with the frequent opening and closing of the iron. The coiled nichrome element is also visible in the photograph, as is the thermostat in the center of the bottom pan. The line cord is usually connected directly to the lower heating element, and the voltage is brought to the upper element by means of two insulated wires that pass through the protective spring.

Servicing a faulty waffle iron involves locating the source of trouble, figuring out how to get at the defective part, and repairing or replacing the part. After the repair the iron should be tested to make sure that it operates properly and that *it is safe to operate.*

If the iron does not heat at all when plugged into a live outlet, the trouble may be due to a defective line cord, a loose connection, a defective thermostat which remains open, a faulty switch, or (most unlikely) both heating elements failing simultaneously. Remove the grids. Loosen the screws that hold the pans in place, and lift the pans holding the heating elements. Using voltage and continuity tests, check the line cord and thermostat. Look for loose connections. Unless the waffle iron was dropped or otherwise shaken up, the trouble is probably in the thermostat or line cord. If it proves to be the line cord, you can fix it by the techniques described in Chapter 3. If the thermostat is at fault, replace it. It is usually held in place by screws. The control knob on the outside is held on by friction and can be pulled off. When the new thermostat is installed, replace the knob so that it turns over the whole scale on the dial. If the switch is bad, replace it.

If one side heats, but the other does not, the trouble is probably a heating element. Check the element which doesn't heat for continuity, and if it is open, replace it. If the upper element doesn't work, the trouble may also be in the wires passing through the spring. Test these for continuity. If one is bad, replace both, since it is easier, and the second wire may be worn and ready to break. When replacing these wires, they are slipped through the spring together, and you should allow four or five inches of slack in the leads to permit flexing without strain.

If the iron doesn't get hot enough, the trouble is probably due to the thermostat opening too soon. Check the line voltage first in this case, since a low line voltage can cause this trouble. If the fault is in the thermostat, replace it.

If the iron gets too hot, the trouble is almost certainly caused by a thermostat that won't open. Replace it.

If the pilot light fails, but the waffle iron works perfectly otherwise, you may ignore the trouble. If you decide to replace the bulb, make sure the new one has the same wattage as the one removed.

Broken feet or handles do not affect the electrical operation, but can be annoying. To get at the screws holding them, it is usually necessary to remove the grids and pans holding the elements. The broken parts can then be replaced. Note carefully the positions of all elements so that they can be replaced correctly.

When you have finished repairing a waffle iron, make a ground check to make sure there is no voltage on the case of the iron. Do this with the plug pushed into the outlet in both positions so that first one and then the other side of the line cord is grounded.

SANDWICH GRILLS

Sandwich grills and special grills to make party snacks are electrically identical to waffle irons and are repaired in the same manner. The only problem may be taking the grill apart. The method of construction varies, but look for screws at the bottoms of deep holes in handles or brackets if the fastenings are not otherwise obvious. Always make a ground check when you finish repairing the grill.

ELECTRIC SKILLETS

An electric skillet or fry-pan has a built-in heating element which is not removable. This heating element should not burn out in normal operation and is unlikely to be damaged

Fig. 5-7   Electric skillet

unless the skillet is dropped with sufficient force to break the connections to the electric contacts. If the heating element is damaged, it is not repairable, and you may as well get a new skillet. The heating element will not be damaged if the skillet is immersed in water, so washing it after use is no problem.

The parts of an electric skillet are shown in Fig. 5-7. The heating element is fastened to the bottom of an aluminum pan. An aluminum sleeve is fastened to the pan in such a manner that the sensing probe on the control unit can be plugged into it. The pan also has a Bakelite handle which is usually removable. The skillet is usually enclosed in a decorative metal housing which encloses the heating element and has a cover to match.

The control unit contains the thermostat. When the control unit is plugged into the receptacle on the skillet and turned on, the line voltage is connected across the heating element. The probe is inserted in the sleeve between the ends of heating element. As the skillet gets hot,

Electric Skillets   **72** the sleeve also gets hot and transfers the heat to the probe. The thermostat is fastened directly to the probe and, at the proper temperature, opens to shut off the electricity. The temperature of operation is adjusted by a small knob which also acts as the on-off switch. A pilot light on the control unit indicates when the skillet is drawing current. *The control unit must never be immersed in water.*

If the skillet does not heat when connected to a live outlet, the trouble may be in the line cord, heating element, switch, or thermostat. As a first check, plug in the line cord, but do not plug the control unit into the skillet. Check the voltage at the output of the control unit by inserting your meter probes into the contacts there. Make sure the switch is turned to the *on* position. If you find voltage at the output of the control box, the trouble is probably the heating element. This can be verified by making a continuity check on the heating element. Note that if both are good, the trouble is in the contacts and these should be cleaned. If the heating element is open, get a new skillet. It is not necessary to get a new control box.

If there is no voltage at the output of the control box, then you can take the box apart and check the line cord, switch, and thermostat separately. Line cord repairs are discussed in Chapter 3. Replace a defective switch or thermostat with an identical replacement part.

If the skillet heats to the wrong temperature, the thermostat may be dirty, corroded, or out of adjustment. Clean the sensing element and all contacts in the control box. If necessary, replace the thermostat. After finishing the repairs always check the case of the skillet for voltage as a safety measure. Do this with the plug in both positions in the outlet.

When washing an electric skillet, be sure to follow the manufacturer's instructions. Disconnect the control box before washing. Do not use harsh or abrasive cleaners when cleansing the unit. Wash the pan as soon as possible after use. Don't soak any part in water and dry everything promptly after washing. Don't store in an oven since the plastic parts can be damaged by heat. If these rules are not adhered to, the surface of the pan may discolor. This

Fig. 5-8   Percolator action

may be esthetically displeasing, but discoloration does not interfere with proper operation of the skillet, and it has no effect on the food.

PERCOLATORS

Electric coffee-makers or percolators are of two types, non-automatic and automatic. The brewing action of both types is the same and is shown in Fig. 5-8. Cold water is placed in the pot and ground coffee in the basket at the

Percolators

top of the overflow tube. The bottom of the overflow tube contains a valve which shuts when pressure builds up inside the tube. This part of the percolator is called the *pump*. Heat is applied to the cold water inside the pump causing it to boil. The pressure closes the valve, and the steam pushes the hot water up the tube, where it is ejected from the top. The water hits a deflector in the lid and falls back onto the basket cover. The perforations in the cover and basket permit the water to ooze through the coffee grounds slowly. After the water has been pushed out of the tube, the pressure inside decreases, and the valve opens allowing more water to enter the pump. The process is then repeated. The electrical portion of the percolator is usually enclosed in a Bakelite housing at the bottom of the appliance. In some types the heater surrounds a well in the bottom of the pot, and the overflow tube and pump are inserted in this well. The principle of operation is the same.

Non-automatic percolators do not shut off automatically. When the user thinks the coffee is done, he pulls the plug. To reheat the coffee or keep it warm, he first removes the basket and tube and then plugs the unit in again. The electrical portion of the pot is thus very simple. It consists of a line cord, a heating element, and a fuse. The fuse prevents the heating element from burning out, if the percolator is left plugged in without any water in it.

If a non-automatic percolator won't heat, the trouble must be in the fuse, the line cord, or the heating element. The Bakelite housing can be removed by loosening a screw or a nut in the center of the bottom of the base, thus exposing the circuitry. The fuse should be checked first since it is the part most likely to fail. Using continuity tests you can determine which part is at fault. For a bad line cord, follow the practices discussed in Chapter 3. If the fuse or heating element has failed, replace it with an identical part. Make sure no part of the electric circuit makes contact with the metal pot before you return the percolator to service.

Automatic percolators shut off by themselves when the coffee is brewed, and then heat is applied to keep the

115V

MAIN HEATING ELEMENT

THERMOSTAT

WARMING ELEMENT

SIGNAL LIGHT

Fig. 5-9   Percolator circuit

coffee warm without percolating further. Although there are almost as many different circuits as there are manufacturers, all are thermostatically controlled, and the principles of operation are similar. Some have a pilot light which goes on when the coffee is brewed. A typical circuit is shown in Fig. 5-9.

During the brewing cycle, the thermostat in Fig. 5-9 is closed, shorting out the signal light and the warming element. The full line voltage is across the main heating element, which heats the water in the pump, causing the percolating action. Although the water inside the pump is boiling, the rest of the water in the pot may still be cool. However, the hot water drips through the coffee grounds and falls back in the pot as coffee, and more cold water enters the pump to be heated. As a result the coffee in the pot gets stronger and hotter. The longer the percolation, the stronger the coffee will be.

When the brewed coffee reaches a predetermined temperature, the thermostat opens. Now the main heating element, the warming element, and the pilot light are in series across the line voltage. The portion of this voltage across the main heating element is small, so that this element no longer gets hot enough to boil the water in the pump. Most of the voltage is across the warming element which warms the bottom of the pot outside of the pump. The pilot light indicates that brewing has ceased, and the warming element keeps the coffee warm. On some models, the temperature at which the thermostat opens can be adjusted by a control on the outside.

Setting the control for stronger coffee simply means raising the temperature at which the thermostat opens, thereby prolonging the brewing action. The same effect can be achieved by starting with colder water. Generally, the thermostat should open at about 190 degrees when the control is set for strong coffee.

As with non-automatic models, the base of an automatic percolator is a Bakelite housing which is held on by a screw or nut in the center. If there is a control to adjust the brewing time, this should be removed before removing the base. Loosen a set screw on the control knob, if there is one; if not, simply pull it off.

If an automatic percolator won't heat at all, the trouble is probably in the line cord or the main heating element. Note that if the warming element failed, the main heating element would still operate. If the thermostat failed in the open position, both heating elements would work, and the water would get warm, although it wouldn't percolate. If the thermostat failed in the closed position, percolation would take place but it would never stop. If the trouble is in the line cord, fix it, referring to Chapter 3. If the main heating element is burnt out, replace it with an identical part. Continuity tests should enable you to determine which part is defective.

Notice that if the warming element or the pilot lamp fails, the percolator will work until the coffee is brewed, but then the pilot lamp will not go on, and the coffee will not be kept warm. Check both the lamp and the warming element for continuity, and replace the defective part. On models without a pilot lamp, if the coffee cools off after brewing, it is due to a defective warming element.

There may also be mechanical failures. If the valve is bent or broken, the percolator will not brew even though the electric circuit is faultless. A bent valve is easily replaced. Other mechanical defects include broken Bakelite base, leaky gaskets, broken handles, and cracked lids. All these parts are obtainable from an authorized service dealer or the manufacturer and are easily replaced. Fig. 5-10

Figure 5-10   Toastmaster model M521 coffeemaker

is an exploded view of the Toastmaster Model M521 Coffee Maker manufactured by the Toastmaster Division of Mc-Graw-Edison Company. Table 5-1 lists the parts in Fig. 5-10. Notice particularly parts number 3, 8, 10, and 15. These are gaskets and washers to prevent leaks. Whenever a mechanical defect requires replacement of a part, you must remember to put in a new sealing device also, since the old one may be deformed during removal of the defective part.

As with all appliances, always check for voltage on the case as a safety measure before returning the percolator to service. Make ground checks with the plug in both positions in the outlet.

ELECTRIC IRONS

There are two types of electric iron: dry and steam. Electrically they are identical, each consisting simply of a line cord, a thermostat, and a heating element. The steam iron has a water chamber in addition to the electrical and mechanical parts of the dry iron. The external parts of an iron are illustrated in Fig. 5-11.

There are a variety of shapes and sizes of heating elements in irons. Nichrome ribbon, nichrome coils, and Calrod are all used. Nichrome ribbon is usually wound on mica sheets and is replaceable if it burns out. The other two types are built into the sole plate, but electrically insulated from it, and cannot be removed. Practically all modern irons use non-replaceable heating elements. This means that when the heating element burns out, the iron should be discarded. Fortunately, with proper usage the heating element will outlast the other electrical parts of the iron.

In Fig. 5-11, a thermostat control is shown on the handle and is located so that it may be moved by the finger tips when holding the iron normally. This control is connected by a rod to the thermostat near the sole plate, and it regulates the opening between the contacts and thus the temperature at which the thermostat opens and closes. The thermostat itself comes in many shapes, but

TABLE 5-1   TOASTMASTER STAINLESS STEEL COFFEE MAKER

MODEL M521

No.	Part
1	top cover knob, black
2	top cover, stainless steel
3	teflon washer (for cover knob screw)
4	machine screw 8-18 x 3/8 (top cover knob)
5	basket cover, stainless steel
6	basket and guide tube assembly, stainless steel
7	percolator tube and pump assembly
8	spout gasket
9	spout mounting screw
10	teflon washer (handle screw)
11	handle screw
12	handle, black
13	sight tube "see level"
14	sight tube reflector, aluminum
15	sight tube seal, rubber
16	body w/percolator element and keep warm element bracket
17	base bracket and stud assembly
18	pivot tension spring
19	nut base bracket, #10-32 x 3/32 x 1/4 hex brass
20	flavor control arm and stud assembly
21	truarc retaining ring
22	hex nut, (terminal pin) 6/32 x 3/32 x 1/4 steel
23	Bakelite base, black
24	terminal pin
25	cord and plug, black
26	signal light lens
27	signal light and resistor assembly
28	flavor control knob
29	flavor control knob screw 3/48 x 1/8 fil. h.ss
30	Bakelite base screw 5/32 x 1/2
31	hole plug 17/32" diameter
32	name plate (on base)
33	nut 8/32" x 1/8 x 11/32" (holds thermostat in place)
34	thermostat assembly
35	3" lead wire, thermostat to percolator element
36	7" lead wire, terminal to thermostat
37	3-5/8" lead wire, percolator element to terminal
38	keep warm element cartridge
39	spout and insert assembly

HANDLE

THERMOSTAT
CONTROL

FILLER HOLE
(STEAM IRONS ONLY)

SOLE PLATE

CORD

COVER
OR
NAMEPLATE

Fig. 5-11   Iron

the principle of all those used in irons is based on the bending of a bimetallic arm.

The line cord and its connections to the rest of the
circuit are the chief sources of trouble in an iron. This is
not surprising, since the cord is continually being flexed
when the iron is in use. Thermostats sometimes burn out or
stick, but thermostat troubles are rarer than mechanical
problems caused by dropping the iron.

When working on any electric iron, always make sure
your work bench is covered with a soft pad or several
layers of cloth. If the bottom of the sole plate is nicked or
scratched, the iron will be useless, even if the electrical
parts work perfectly.

Fixing an iron is very simple once you have figured
out how to take it apart. Each model is a puzzle, but once
the key is found, the iron can be disassembled easily. Never
try to force it apart. On many models, a flexible metal
cover, sometimes used as a nameplate, conceals a large
nut in the center of the iron. This cover is shown in
Fig. 5-11. It can be pried off gently with a screwdriver
exposing the nut. You then remove the nut, and the
handle can be lifted off, giving access to the electrical
circuit. The temperature scale may have to be pried forward and off on some models before the handle can be
lifted from the rest of the iron. On steam irons, the filler
tube should be pulled forward and out before lifting off

Electric Irons

the handle. In order to be sure that you reassemble the iron with the thermostat control in the same position as when you disassembled it, you should push this control to one extreme end before disassembling. Then when you reassemble the iron, make sure the control is at the same end.

The cord is usually attached to the rest of the circuit by two spade lugs which are clamped under their respective screwheads. To remove the cord, loosen the screws, and the cord can be pulled out. It is not necessary to remove these screws completely.

At least one model is held together by a plastic pin located at the top of the handle near the front. To take this iron apart, drive this pin into the iron until the handle is loose. The pin can be recovered after the iron is taken apart and should be replaced in its original position when the iron is reassembled.

In some models, it is necessary to pry off the indicator arrow at the front of the handle and to remove the control knob and adjusting screw. This will reveal a square hole at the front of the handle. You should insert a Phillips head screwdriver into this hole and loosen the screw at the bottom, which holds the iron together.

After you have the iron apart, it is a simple matter to make voltage tests and continuity tests to determine the trouble. With the plug inserted in a live receptacle, you should find voltage across the terminal screws of the line cord. Check this voltage if you have a meter, since an iron works poorly if the line voltage drops as little as 10 percent. However, in most locations, a line voltage drop is rare.

If the iron won't heat, the trouble is in the line cord, heating element, or thermostat. There is nothing else. If there is no voltage at the line cord terminals inside the iron, the line cord is at fault. Fix it, using the techniques discussed in Chapter 3. If the line cord is all right, make continuity checks on the thermostat and heating element. Usually the thermostat can be replaced if it is bad, but the heating element cannot be replaced. Note carefully how the thermostat is connected, and if it needs replacement, use an identical part.

If the iron gets warm, but not hot enough to iron, the trouble may be a low line voltage (possible, but unlikely) or a thermostat that is loose or shuts off too soon. Put a clip lead across the thermostat, bypassing it. If the iron now gets hot, the trouble is in the thermostat. The contacts may be dirty or bent. Generally, it is best to replace the unit rather than try to fix it.

A steam iron is identical to a dry iron electrically, but may succumb to troubles caused in the water chamber. All manufacturers recommend using distilled water in their steam irons, but some users forget or need to iron when no distilled water is available. If tap water is used, impurities cause deposits inside the water reservoir in the iron, and in the valve and escape holes. When these become clogged, the iron can still be used as a dry iron, but not as a steam iron. To clean out the deposits, soak the unit for at least 12 hours in vinegar. The deposits can then be flushed out with water. A steam iron may also develop leaks. Take the iron apart and replace any gaskets or sealing washers which show signs of deterioration.

Before reassembling an iron make sure that no stray strands or wire will contact the case of the iron when it is put back together. After assembly check the iron for voltage on the case. This ground check should be performed for both positions of the plug in the receptacle.

TOASTERS

Although there are many different types of electric toasters on the market, to most people an electric toaster is the popular pop-up model. Slices of bread are dropped into slots in the top of the toaster and are moved down into the body of the appliance either manually or automatically. When the bread is toasted, it pops up, and the toaster shuts itself off. Other types, resembling sandwich grills, table ovens, and the like, are relatively simple devices and are repaired by the techniques listed under waffle irons or hotplates. The automatic pop-up toaster is *not* a simple device to repair. Both the electrical circuit and the mechanical operation are more complicated than in most other small electrical appliances.

To make matters more complicated, there are many different kinds of pop-up toasters operating on widely different principles. If you remove the case from a toaster while it is still in operating condition, you might be able to figure out its method of operation by watching it work. But if it is out of order, this becomes a difficult task. Fortunately, all toasters give many years of trouble-free service, and when they do go wrong, their owners usually welcome the excuse to buy the latest improved model. Nevertheless, some repairs are relatively simple, and you can do them if you do not let yourself be awed by the apparently complicated machinery.

The first pop-up toasters were timed by a spring-wound motor. When the lever was pushed down, it wound the timing motor and turned on the switch. Then when the motor ran down, the electricity was turned off and the latch holding the carriage down was released. The carriage snapped up bringing the toast partially out of the toaster. In order to prevent the toast from flying out completely, a pneumatic stop, similar to that used to prevent screen doors from slamming, cushioned the movement of the carriage. A color control knob on the toaster regulated the time that the carriage stayed down and the electricity remained on. At any one setting of the knob, the duration of the toasting cycle was constant. Therefore, the first slice, put in when the toaster was cold, would not be as dark as subsequent slices of toast. The user was advised to preheat the toaster by letting it operate through one cycle before putting in the first slice.

Preheating the toaster was a slight nuisance, and naturally manufacturers devised better toasters which eliminated this problem. Although many of the old clock-timed toasters are still in use, they haven't been manufactured for years, and you should not concern yourself with how to repair them. However, if you know how to fix a modern pop-up toaster, the old clock-timed model will pose no problems.

Fig. 5-12 Clock-timer with compensator

One type of automatic toaster uses a bimetallic regulator or compensator along with the clock mechanism. The regulating portion of the toaster is shown in Fig. 5-12. When the toaster is cold, the bimetallic arm is straight and has no effect on the timing apparatus. As the temperature rises inside the toaster, the bimetallic arm bends toward a lever which regulates the speed of the clock. As the temperature increases further, the bimetallic arm pushes the speed-controlling lever and causes the clock to run faster, shortening the toasting time. Thus, the second and subsequent slices of toast are no darker than the first. The original position of the speed-controlling level is set by a control knob on the case to choose light to dark toast.

Note that a *bimetallic arm* is used as a compensating device, but it is not called a thermostat. A thermostat also uses a bimetallic arm, but strictly speaking a thermostat is a switch which is turned off and on by heat.

In another type of automatic toaster using a similar speed-regulating control, the bimetallic arm is in contact with the bread to be toasted. If the bread is fresh, it contains more moisture and takes longer to reach a toasting temperature. The bimetallic arm remains straight for a longer period of time. On the other hand, if stale dry bread is used, it gets hot quickly, and the bimetallic arm bends earlier to speed up the clock. This type of toaster not only maintains constant color for later slices of toast, but it does so for stale and fresh bread, also.

In both types of toaster described above, when the clock runs down or stops, some sort of mechanical device shuts off the electricity and moves the latch holding the carriage down. The carriage pops up, cushioned by the shock absorber, so that the toast can be removed. The toaster is then ready for the next cycle.

One type of toaster uses a *hot-wire carriage release.* A separate strand of wire is connected across the line voltage. Current flows through the wire causing it to get hot and expand. The wire holds the carriage down by means of a mechanical linkage. When the thermostat shuts off the current, the wire cools and contracts, releasing the carriage so that it pops up.

Another type of automatic toaster uses a two-stage thermostatic control. This toaster has an auxiliary heating element wrapped around the bimetallic arm. A schematic diagram of the electric circuit is shown in Fig. 5-13. When the carriage is lowered, the main switch is closed, causing current to flow through both the auxiliary heater and the main heating elements. The heat from the auxiliary heater causes the bimetallic arm to bend. This is the first stage. The bent arm causes the auxiliary switch to close. Referring to Fig. 5-13, you can see that this shorts out the auxiliary heater from the circuit, but current still flows through the main heating elements. In this second or cooling stage, the bimetallic arm tries to straighten. However, at the same time that the arm closed the auxiliary switch, a mechanical device gripped the arm in such a way that it could not return to its original position. Then in trying to straighten, one end moves toward a release which trips the carriage, permitting it to pop up. When the carriage moves up, it opens both switches and releases the bimetallic arm, thus resetting the toaster for the next slice of bread.

The mechanical device which detains the bimetallic arm in the two-stage circuit takes many forms. One possible configuration is shown in Fig. 5-14. In (a), the bimetallic

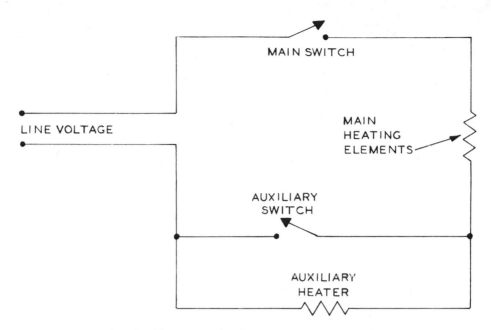

Fig. 5-13 Two-stage circuit

arm is horizontal and straight. In (b), the arm has bent because of the heat and a mechanical detent has moved under the arm. Now when the arm cools, the detent prevents the arm from returning to its original position (shown by the dotted lines). Thus it straightens at an angle to the horizontal, and its free end trips a lever freeing the carriage. The detent is released when the carriage moves up, and the bimetallic arm then moves back to the position shown in (a).

In the first two-stage toasters, the auxiliary heating element was an electrically insulated coil slipped over the bimetallic arm but not fastened to the arm otherwise. In later models, the heating element is firmly fixed in place, so that it is always located at the same position on the bimetallic arm.

The heating elements in all pop-up toasters are made of flat nichrome wire wound back and forth close to the positions that the bread will occupy. These heating elements are *not insulated* electrically and it is dangerous to touch them when the toaster is plugged in, even if the

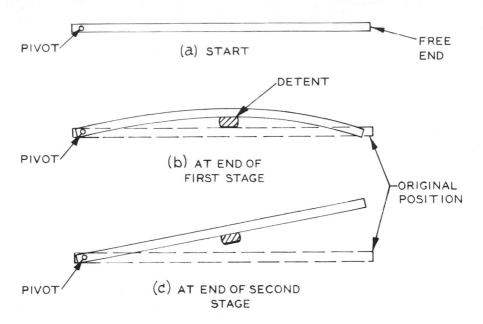

PIVOT — (a) START — FREE END

DETENT

PIVOT — (b) AT END OF FIRST STAGE

ORIGINAL POSITION

PIVOT — (c) AT END OF SECOND STAGE

Fig. 5-14 Mechanical hookup of two-stage toaster

toaster is off. This is explained in Fig. 5-15, which depicts the main circuit in the toaster. When the toaster is plugged into an outlet, one side of the heater circuit is connected directly to the outlet. This could be the high side of the line and you are unlikely to know which side of the plug is connected to the hot side. Now if you touch a heating element and ground at the same time, you will be across the line voltage. When working on a toaster always make sure it is disconnected except when you are making voltage tests.

In Fig. 5-15, four heating elements are shown, which is usual for a two-slice toaster. These four elements are *not* identical, and thus if you have to replace one, make sure the new element has the same power or current rating as the old one. For example, in a 1200-watt toaster, the two outside elements may be rated at 315 watts each, and each of the inner ones at 285 watts. When slices of bread are in the toaster, the outer surfaces then are exposed to 315 watts. Each inner surface receives 285 watts from its adjacent element, but is also "sees" the other inner element

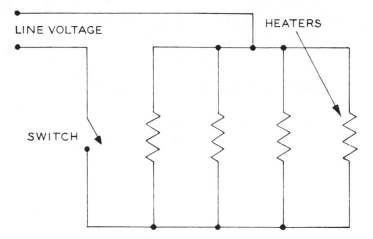

Fig. 5-15 Heating elements

and receives some heat from that. In practice, it receives about 10 per cent of the heat from the other element, so that in this example it would receive about 314 watts on each inner surface. Thus, all four surfaces receive about the same amount of heat. When ordering replacement elements, you can specify either current rating or power rating, but the best thing is to specify the model number of the toaster and whether it is an inside or outside element.

If an automatic toaster doesn't work at all, fixing it is relatively simple. The electrical portion consists of a line cord, a switch, and the heating elements. Since it is unlikely that all four heating elements would burn out simultaneously, the trouble must be in the outlet, the line cord, or the switch. The fault might also be due to an open circuit caused by a broken wire or a loose connection. First make sure that the outlet is live by testing for voltage. With the toaster disconnected from the line, remove the cover. The control knob usually slips right off since it is held on by friction. The handle which moves the lever is usually held on by a set-screw. The control knob and lever handle should be removed before removing the screws which hold the cover. It may also be necessary to remove some decorative trim.

With the cover removed, set the toaster right side up and plug it in. Check the toaster side of the line cord for voltage. If the line cord is bad, fix it or replace it, using the techniques discussed in Chapter 3.

If the line cord is all right, make a visual check of the toaster for loose connections or broken wires. Assuming everything appears normal, depress the carriage, and make a voltage check at both sides of the switch. You might also remove the plug, and with the carriage depressed, check the switch for continuity. A faulty switch should be replaced by an identical part.

If one of the heating elements fails to light, it may be open, but also it may only be disconnected. This is simple to check. A faulty heating element should be replaced with an identical part.

If the electrical portion of the toaster operates satisfactorily, but it won't pop up, or it pops up too soon, or otherwise acts improperly, the trouble is mechanical. Frequently, mechanical troubles can be cleared up simply by cleaning the crumbs out of the toaster. Many of the levers and linkages in the toaster depend on gravity for their action and require loose, clean joints to operate. Crumbs in a joint can interfere with proper operation. If necessary, put a drop of light machine oil on any sticky joint.

If the auxiliary coil on the bimetallic arm burns out, the carriage will not pop up when the toast is done. Check this heater for continuity. If it is open, replace the coil, if it can be removed, or the whole bimetallic assembly otherwise.

If toast pops out of the toaster, the shock absorber is probably defective. It should be replaced. A temporary repair can be made by stretching the spring that lifts the carriage so that it doesn't pull with as great a force. This is not recommended, however, since then the spring will not have enough force after the shock absorber is replaced.

To replace a heating element, it may be necessary to straighten the small crimped ears that hold it in place. This must be done carefully with long-nosed pliers, making sure that each ear is moved only enough to remove the old element. Fortunately this is something you will not have to do often, for too frequent bending would cause the ears to break off.

Before attempting extensive repair work, note carefully how the bread guide wires are mounted in the toaster. These thin wires keep the bread slices in the proper positions relative to the heating elements, but because they are easily bent, they can be damaged by careless handling. If you observe how they are held in the toaster, you can replace or straighten them correctly when they are damaged.

Ornamental trim has no effect on the operation of the toaster but can be damaged if the toaster is dropped. (Internal parts also will be damaged.) The trim is easily replaced since it is usually held on by screws in an obvious manner.

When you have finished repairing a toaster, always make a ground check before returning it to service. Make sure no voltage exists between the case and ground regardless of the position of the plug in the outlet.

ROTISSERIES

A rotisserie or roaster is essentially a portable oven and in principle of operation is identical with the large wall ovens used in many households. It is basically an enclosed box with a heating element and a revolving spit, but more deluxe models come with many additional features. These include a separate baking element, a surface heater, and a warming compartment. Some models with only one heating element may be turned upside down so that they can broil food when the heating element is at the top or bake food when it is at the bottom. Many models have an auxiliary timer which shuts off the rotisserie at a preset time. Some also include a timer which turns on the switch at a preset time as well as turning it off.

The broiling element in a rotisserie is usually a Calrod heater mounted in the top of the heating chamber, although coiled nichrome wire is also used. If a separate baking element is available, this is usually removable and may be plugged into a socket near the floor of the rotisserie. The baking element, also, may be Calrod or nichrome coils. The heating circuit, besides the heating element and line cord, usually includes a switch, a pilot light, and some sort of heat control such as a thermostat.

The spit is rotated by a small electric motor. A separate switch is provided to run this motor, so that the spit may or may not be rotated when the broiler is used. The spit is also removable, and when it is out, as for baking or roasting, the motor can remain turned off. However, in some models, the motor runs constantly and through a cam, controls the temperature both when it is driving the spit and when the spit is removed.

The main problem in servicing a rotisserie is in isolating the circuit you wish to test. Thus, you might make a continuity check across the prongs of the line cord, but you would be checking several circuits in parallel, and if any one of these is all right, you would have continuity. The motor circuit, the broiling element, and the baking element must be checked separately.

A little thought will indicate what part of the circuit is at fault. If nothing works, the trouble is in the line cord, the main switch, or the house circuits. Since a rotisserie draws more current than the average appliance, especially when both heating elements and the motor are turned on simultaneously, it may easily overload the house circuit, when other electrical appliances are also in operation. Hence, if a rotisserie doesn't work at all, check the outlet first. If the outlet is live, check the line cord, using the techniques of Chapter 3, and if the line cord is at fault, fix it by the methods discussed in that chapter. Check the switch for continuity. Replace it if necessary with a similar or one which has an equivalent current rating.

If everything works but the pilot light, you might ignore the trouble, and at your leisure replace the bulb at some later date. The bulb has no effect on the operation of the rotisserie.

If the motor turns, but either heating element fails to come on, the trouble is not in the line cord. It may be in the heating element itself or it may be due to a defective heat control unit. Check the heating element for continuity, making sure nothing is connected in parallel with it. If the heat control unit is a thermostat, make a continuity check across it, too. It may be stuck in the open position. Also check the connections to the heating elements. Sometimes, especially on plug-in elements, the contact pins oxidize, and this layer of metal oxide acts as an insulator. The contact pins should be cleaned with sandpaper. Never use emery paper to clean electric contacts, since the dust from this paper is conductive, and if it falls in the wrong place, it can cause a short circuit.

If the elements heat up, but the motor doesn't turn, check the motor switch. If the switch is all right and there is voltage across the motor terminals, the trouble is in the motor. The operation of electric motors and repairs you can make are discussed fully in the next chapter.

In one type of rotisserie heat control is maintained by a cam on the motor. The cam opens and closes an auxiliary switch to the broiling element. When the control knob is set at *high,* this switch is out of reach of the cam, and the heating element is on all the time. At *medium,* the switch is pushed closer, and the cam opens it half the time. Thus, the heating element is drawing current only half the time and gives off half as much heat. The control knob at lower and intermediate settings moves the switch to an appropriate position so that the cam can open and close it to produce the desired heat. This type of heat control is used only on the broiler. An additional heat control in the form of a thermostat may be used for baking or roasting. If the cam slips on its axle, the switch may be stuck in the open position, or the cam may never open it, so that it remains closed and gives too much heat. A slipping cam is easily detected by observation and easily corrected by tightening the set-screw holding the cam to the axle.

After repairing a rotisserie, make a ground check before using it again. There must be no voltage on the case of the rotisserie with the line cord plugged into the outlet in any position.

HEATERS

Electric heaters come in a wide assortment of shapes and sizes. Small, portable units are used for local heating or

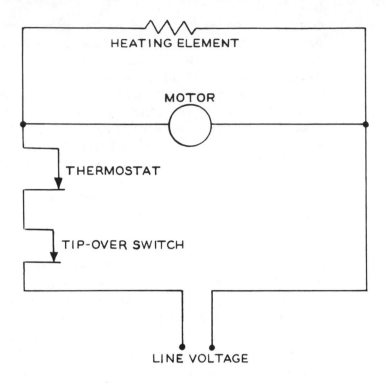

HEATING ELEMENT

MOTOR

THERMOSTAT

TIP-OVER SWITCH

LINE VOLTAGE

Fig. 5-16   Simple heater circuit

auxiliary heating. Larger units may be installed permanently, in baseboards for example, as prime heaters. Although circuits differ somewhat, the principles of operation of all are similar.

The heating circuit consists of one or more heating elements, a thermostat, selection switches to choose the desired amount of heat, and a tip-over switch. In addition there is a fan to blow the heated air into the room to be heated. The fan is operated by a small AC motor.

A simple circuit for a heater is shown in Fig. 5-16. When the plug is inserted in a wall receptacle, the line voltage is applied across both the heating element and the motor through the tip-over switch and the thermostat. The temperature at which the thermostat closes is controlled by a knob on the front of the heater. In some models there is no on-off switch, and with the thermostat at its lowest setting the heater will be turned on by the

Heaters

HEATING ELEMENT

MOTOR

VOLTAGE DROPPING RESISTORS

SAFETY THERMOSTAT

TIP-OVER SWITCH

HIGH MEDIUM LOW

LINE VOLTAGE

Fig. 5-17   Pushbutton variable heat control

thermostat whenever the ambient temperature (temperature in the vicinity) drops to about 55 degrees. To make sure the heater remains off when it is not in use, the wall plug should be disconnected.

A tip-over switch is an important part of every portable heater. This is a pushbutton switch on the bottom of the heater which must be held in for the heater to operate. With the heater in its normal, upright position, the tip-over switch is pushed in because it is against the floor. If the heater is picked up or tipped over, a spring forces the pushbutton out, disconnecting the electricity. The main purpose of this switch is to prevent fires.

The circuit of Fig. 5-16 produces only one level of heat. When the room or space is sufficiently warmed, the thermostat shuts off the heater; and when the room cools down, the thermostat turns it on again. Other types of circuits are shown in Figs. 5-17 and 5-18. In the heater shown in Fig. 5-17, only one element is used, but the voltage across it is changed by a series of switches. Pushbutton switches are shown in the figure, but any type

Heaters

Fig. 5-18 Two-element heater

could be used. When the high, medium, or low button is pushed, the motor is also   connected at the same time. The dotted line around the switch in the motor line indicates that it is not necessary for the user to close this switch since it is done automatically when any of the others are closed. When the *high* switch is closed, the full line voltage appears across the heating element, producing maximum heat. When the *low* switch is closed, two resistors are placed in series with the heating element. The voltage across the heating element is reduced, since there is a voltage drop across the resistors. This causes a lower current through the heating element and consequently lower power or less heat. At the *medium* setting, only one of the voltage dropping resistors is in the circuit, so that the heat produced is intermediate between the other two. The pushbuttons are arranged so that pushing down one releases all others. An additional button is used for *off* which disconnects all the rest, including the motor switch.

The safety thermostat shown in Figs. 5-17 and 5-18 is not used to control the operation of the heater as was the thermostat in Fig. 5-16. Instead this is strictly a safety device which shuts off the heater completely whenever the temperature gets dangerously high.

The heater depicted in Fig. 5-18 uses two heating elements. Typically, one may be rated 400 watts and the other 800 watts. Separate switches are shown, although one multi-purpose rotary switch may be used. When the switch is set for *low,* only the 400-watt unit is in the circuit; at *medium,* only the 800-watt unit is connected; and at *high,* both are connected, producing 1200 watts. Thus, three levels of heat are available.

It is also possible to produce three levels of heat using two identical elements. A typical circuit is shown in Fig. 5-19. Two elements, each with a resistance of 23 ohms are used. When the switch is set to *high,* both elements are connected in parallel, as in (a). Since there are 115 volts across each element, the current through each is five amperes. (Recall that voltage equals current times resistance.) Thus, the total current is 10 amperes, and the power is current times voltage or 1150 watts. When the switch is set for *low,* the elements are connected in series, as in (b). Now there are 46 ohms across the 115-volt line. The current is 115/46 or 2.5 amperes. The power is, therefore, 115 times 2.5 or 287.5 watts. In the medium position only one element is across the line, as in (c). The other is disconnected. The current through this element is five amperes, and the power is thus 5 times 115 or 575 watts. This type of circuit is used in some rotisseries as well as in some electric heaters.

If neither the fan nor the heating element operates when the heater is plugged into a live outlet, the trouble may be in the line cord, the thermostat, one of the switches, or a loose connection. Check the line cord first. Usually, the back of the heater can be removed by loosening four screws, and this permits access to the electric circuit. If there is no voltage at the heater end of the line cord, repair or replace the line cord as described in Chapter 3. If the line cord is in good shape, as indicated by voltage at the heater, make continuity checks on the switches and thermostats. When checking the tip-over switch, you must

Heaters

(a) HIGH, 1150 WATTS

(b) LOW, 287.5 WATTS

(c) MEDIUM, 575 WATTS

Fig. 5-19 Heating circuits

make sure the switch is depressed when you check for continuity. You can short out the switches and thermostat temporarily using clip-leads. Now the heater should operate if one of these is bad. Remove one clip-lead at a time, and when you have removed the lead across the faulty component, the heater will no longer work. Be sure you pull out the plug whenever you put on or remove a clip-lead.

If one or more heating elements fails to work, although the fan motor runs, the trouble may be in the element or the switch connected to it. In either case, a continuity check will identify the defective part. Defective switches and thermostats should be replaced with parts with similar ratings, if possible, although slight deviations will not impair the operation of the heater. In some heaters, individual heating elements may be removed and replaced with new ones. In others, the heating elements and reflector are not separable but must be replaced as a unit. These parts are generally available at authorized service dealers, but if not, they may be obtained from the factory at the address on the nameplate.

If a heater seems to work but doesn't produce enough heat, one of the heating elements is probably defective. The trouble may also be in the switch connecting this element to the line. Continuity checks will enable you to locate the trouble.

Heaters

If the heating elements work, but the blower motor is inoperative, the trouble may be a defective motor, but it may also be mechanical, such as a bent fan blade jammed against the frame of the heater. First make sure that the motor and fan are free to rotate without obstruction. Make continuity checks on the switches which operate the motor, and also check to see that all connections are in order. If you decide that the motor is at fault, it may be possible to repair it, or if not, replace it. Motors are discussed in detail in Chapter 6.

Finally after repairing the heater, make a ground check to be sure there is no voltage on the case for any position of the plug in the socket.

VAPORIZERS

An electric vaporizer is one of the simplest and most efficient of all electrical heating appliances. *All* the electrical energy is converted into heat, and none of this heat is wasted. It is all used to heat water. This is accomplished by making the water to be heated a part of the electric circuit. This is shown in Fig. 5-20. The line cord is connected to two electrodes which are close together, but not touching, at the bottom of an insulated bowl. When this bowl is empty, no current flows. However, when water is placed in the bowl, the water completes the circuit, and current flows through the water. Since water has some resistance, the electric current produces heat, just as it does in any resistance. But this heat is entirely in the water, so none is wasted. Note that when the water boils, the circuit of Fig. 5-20 will shut itself off as soon as all the water is evaporated.

In a practical vaporizer there must be some way of replenishing the water supply as the water boils away. One method of doing this is shown schematically in Fig. 5-21. A large jar acts as a water resevoir. It is filled with water and covered with a cap in which there is a small hole, about one quarter inch in diameter. This jar is inverted and supported in the insulated base so that the water will run out

Fig. 5-20   Electric circuit of vaporizer

through the hole into the well containing the electrodes. When the water level in the bottom of the bowl reaches the cap of the jar, no more water will come out of the reservoir. This happens because the pressure of the atmosphere pushing on the water in the bowl is greater than the pressure of the column of water left in the jar. Now as the water is heated and evaporates, the level in the bowl goes down below the cap of the jar, and then more water comes out to keep the level constant. This continues until all the water is boiled away.

The electrical part of a vaporizer is amazingly simple, and almost nothing can go wrong with it. Check the line cord, and if that's all right, the circuit is all right. The problems that arise are not basically electrical in nature.

In some applications, medication is placed in the bottom of the bowl to be evaporated along with the water. If the medication is greasy, it can leave a coating on the electrodes, and when this coating is thick enough, it may act as an insulator to prevent the electrode from making electrical contact with the water. To prevent this or to correct it when it does happen, the vaporizer should be washed in soapy water after each use.

If the local water supply contains a high level of mineral salts, these may be deposited on the electrodes as water boils off. Usually, these minerals conduct electricity

Vaporizers

Fig. 5-21 Water reservoir

and do not interfere with the operation of the vaporizer. However, eventually they may build up to a point where they short circuit the electrodes, usually blowing a line fuse. To prevent this, you should scrape the mineral deposit off the electrodes from time to time.

A special application of this method of heating water is a warmer for baby's bottles. A measured amount of water is placed in the bowl containing the electrodes, and the bottle is also inserted above the water. The water is heated and boils, and the steam heats the bottle. Since the amount of water is controlled, the time for it to boil away is also controlled. This time is selected to bring the baby's bottle to just the right temperature.

Chapter 6

Electric Motors

The electrical appliances which are discussed in Chapter 5 are primarily devices for producing heat. To be sure, the rotisserie has a motor which turns a spit so that the food is exposed to the heat evenly, and the electric heater has a motor which rotates a fan to blow heat into the room, but the prime purpose of both of these appliances is producing heat. On the other hand, in many appliances, the motor is the important element, and the motion imparted by the motor is used to accomplish other household tasks. Motors are used to drive compressors in refrigerators and to produce vacuum in vacuum cleaners, to mix food in mixers and blenders, and to rotate clothes in washers and dryers. Look around your house and you will probably be surprised at the large number of electric motors it contains.

Motors are rated according to the amount of power they deliver. Since small motors are usually somewhat less than 70 per cent efficient, it is necessary to put in 100 watts to get out the equivalent of 70 watts in mechanical energy or turning force. Motor ratings are usually given in *horsepower,* and one horsepower is equal to 746 watts. Thus, a 1/4-horsepower motor *puts out* the equivalent of 1/4 of 746 or 186.5 watts. Since this is 70 per cent of the input, the motor requires about 266 watts to drive it. Thus, it will draw about 2.3 amperes from a 115-volt line. (Watts equals volts times amperes.) On some appliances with motors, the nameplate may list the power or the current instead of the horsepower of the motor. A blender, for example, may show 700 watts on its nameplate. Using the 70 per cent efficiency rule, the output of the blender motor would be 490 watts, which is about 2/3 of 746. The motor in this blender is a 2/3-horsepower motor. As another example a small fan may draw one ampere. To determine the horsepower of the fan motor, you must first figure out the power fed in, which is the product of current and voltage. In this case, one ampere times 115 volts gives 115 watts. Assuming 70 per cent efficiency, the output is about 80 watts, which is slightly more than 10 per cent of 746. This motor would be called a 1/10-horsepower motor.

Motors that put out less than one horsepower are called *fractional-horsepower motors.* All the motors used in home appliances are fractional-horsepower, ranging from about 1/20 to 3/4 of a horsepower.

Motors may be designed to run on alternating current only, on direct current only, or on either. A motor which will operate on either AC or DC is called a *universal* motor. *Induction* motors run on AC only. Most home appliances use either a universal motor or one of two types of induction motor, a *split-phase* motor or a *capacitor-start* motor. A third type of induction motor, the *shaded-pole motor,* is occasionally used in very small fans.

In selecting a motor for a particular appliance, a manufacturer must consider the properties of the motors and the requirements of the appliance, as well as cost. A motor which might work well in an electric heater would not be satisfactory in a blender. On the other hand, a blender motor could work in a heater, but would be more expensive than the motor actually used there.

Criteria for selection include speed, reversibility, and torque. Should the speed be constant or variable? Can the direction of rotation be reversed? Can the motor start under a heavy load? Is the speed of rotation affected by the load? A clock motor must run at constant speed. Many other appliances have constant speed motors, also, although it is not as important as in a clock. Some appliances have two or more speeds. The speed may be continuously variable as in some mixers, or it may be possible to switch from one to another constant speed as in some fans or blenders. Appliances generally do not need reversible motors, although some motors can be reversed by switching connections on the windings. Some models of garbage disposers do have a switch for reversing direction of rotation, so that both sides of the cutting teeth can be utilized.

The *torque* is the turning force of the motor. If the motor can start under heavy load, it is said to have a *high starting torque.* Once the motor is running, it requires less force to maintain rotation. If an increase in load causes the motor to slow down, it is said to have *poor regulation.* Most orange juicers, for example, will slow down if you exert too much pressure on the orange.

Another important consideration is *starting current.* When a motor is turned on, there is a surge of current until the rotor starts to rotate. This is technically called *locked-rotor current* since it is the current with the rotor stationary. When the rotor is turning, the current drawn by the motor is much less than the locked-rotor current. A high starting current may cause lights to flicker and line voltage to decrease temporarily. With appliances which are started infrequently, the occasional flicker is tolerable, but in a refrigerator or air conditioner, which turns on and off frequently, a high starting current would be a constant source of annoyance.

BASIC INDUCTION MOTOR

A motor consists of two main parts: the *stator* or stationary part, and the *rotor* or rotating part. The stator is connected to a frame for mounting. The rotor should rotate freely and is usually supported in bearings mounted in the same frame.

In a basic induction motor, the stator is a coil of wire, surrounding the rotor, but *not* in contact with it. Current through the coil creates a magnetic field which then induces current in the rotor. The rotor current also creates a magnetic field which tries to line up with the magnetic field in the coil. Since alternating current is used, the magnetic field in the coil rotates, and thus the rotor should rotate, also.

There is only one difficulty with this basic motor. It won't start. If you spin the rotor after switching on the current in the coil, it would continue to rotate at constant speed as long as current flowed. It would rotate in either direction, but something must be added to start the motor. The different types of starting devices lend their names to the different types of motors.

SPLIT-PHASE MOTOR

A split-phase motor is a practical induction motor which uses a second coil as a starting device. The stator coils are called field coils since they set up magnetic fields. In the split-phase motor, the main coil is called the *running* coil or running winding, and the second coil is the *starting* coil. Their respective magnetic fields are the running field and the starting field. The two coils are displaced magnetically, and the two-phase field causes the rotor to start spinning. After the motor has started, the starting coil is disconnected by a centrifugal switch.

The rotor of an induction motor is a solid casting with no coil windings on it. It is referred to as a *squirrel cage rotor,* and is illustrated in Fig. 6-1. The ends are supported in bearings in the frame.

The stator of an induction motor could be simply a coil, but it would be inefficient. In practice it consists of pole pieces made of laminated metal strips on which the

Fig. 6-1   Squirrel cage rotor

field coils and starting coils are wound as shown in Fig. 6-2. The poles for the starting coil are displaced from the running poles as shown in the figure. The current in the starting coils is out of phase with that in the running coil so that a rotating magnetic filed is produced to start the motor. The pole pieces are joined by a laminated metal ring to close the magnetic field.

After the motor is running, the starting coil should be disconnected since it is no longer needed. If the starting coil were not disconnected, the motor would still run, but it would draw current in the starting coil which would increase the power put into the motor, but would not increase the power out. Thus, the motor would be inefficient and would tend to overheat. In practice, a centrifugal switch is used to disconnect the starting coil. This switch is mounted on the rotor and is held closed by a spring. When the motor attains sufficient speed, centrifugal force overcomes the pull of the spring and forces the switch open. The switch is connected in series with the starting coil through slip-rings on the rotor.

Electrically, the circuit of a split-phase motor is as shown in Fig. 6-3. In (a), both the running coil and the starting coil are connected across the 115-volt line. This is the position at rest or at the start. When the motor is running, the centrifugal switch is open, and thus the starting coil is disconnected from the 115-volt line, as in (b). The motor continues to run since the running coil is still connected. If the main switch connecting the motor

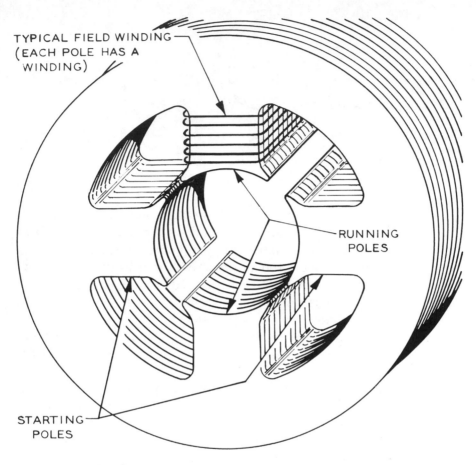

TYPICAL FIELD WINDING
(EACH POLE HAS A
WINDING)

RUNNING
POLES

STARTING
POLES

Fig. 6-2   Pole pieces

to the house line is disconnected, the motor starts to slow
down and eventually stops. Even before it has reached a
complete stop, the spring on the centrifugal switch will
overcome the decreasing torque and close the switch,
returning the circuit to that shown in Fig. 6-3(a). Note
that if you made a continuity check at the line input to
the motor, you would have continuity even if one of the
coils were open, since both are in the circuit when the
motor is at rest.

In Fig. 6-3, the rotor is indicated by a circle with
smaller circles inside of it. This is the standard symbol for

Fig. 6-3   Split-phase motor

a squirrel cage rotor. Note that there are no electrical connections to the rotor in a split-phase motor. The electrical connections to the centrifugal switch, which is indeed mounted on the rotor, are not considered connections to the rotor. The rotor is usually held in place by bearings in end bells on the motor case. Frequently, long screws passing through the entire motor body join the end bells together and thus hold the motor together as indicated in Fig. 6-4. The rotor is not a snug fit since it must rotate freely. Endplay should be at least 0.005 inch, and even 5 or 6 times this amount is permissible.

A split-phase motor has a high starting current and low starting torque. However, because this type of motor is relatively inexpensive, it is used for washers, dryers, and some fans. Speed of rotation is constant. The direction of rotation may be reversed by changing the coil connections when the motor is at rest.

CAPACITOR-START MOTOR

In the split-phase motor it is necessary to have two out-of-

Capacitor-Start Motor

109

ROTOR

END BELLS

ASSEMBLY SCREWS

MOTOR BODY CONTAINING
FIELD COILS

Fig. 6-4   Physical construction

phase fields to start the motor. This is accomplished by winding the two field coils differently. Another way of producing a different field in a second coil is by placing that coil in series with a capacitor. Schematically this is shown in Fig. 6-5. Coil A is directly across the line. Coil B is in series with capacitor C. The capacitor introduces a phase shift across coil B, so that the two coils are out of phase.

This principal is used in a capacitor-start motor as shown in Fig. 6-6. In (a), with the motor at rest, the circuit is essentially that of Fig. 6-5. The different phase in the starting coil is produced by the capacitor. When the motor is running, the centrifugal switch opens, as in Fig. 6-6(b). Now the capacitor and the starting coil are removed from the circuit. Note the similarities between Figs. 6-5 and 6-6.

The capacitor in a capacitor-start motor is mounted

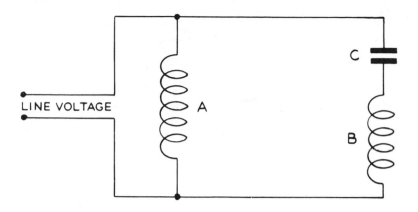

Fig. 6-5   Capacitor phase shift

outside the main motor casing. The external bulge easily identifies this type of motor. A typical shape is shown in Fig. 6-7.

The capacitor on a capacitor-start motor causes an increase in starting torque (as compared to a split-phase motor) and a decrease in starting current. Thus, this motor can be used in refrigerators and air conditioners, since it

Fig. 6-6   Capacitor-start motor

Fig. 6-7   Capacitor bulge

does not cause objectionable light flickering. The capacitor represents an additional expense, so this motor is more expensive than the split-phase motor for the same power rating. The capacitor-start motor runs at constant speed, and direction of rotation can be changed by changing the coil connection when the motor is at rest.

SHADED-POLE MOTOR

Another method of producing the required phase difference is by *shading* the field pieces. This means simply slotting the field poles near one end and wrapping a copper strap around the "shaded" section, as shown in Fig. 6-8. A current induced in the copper strap produces the required out-of-phase magnetic field and starts the motor. This motor differs from the others mentioned above in that the

Shaded-Pole Motor   **112**

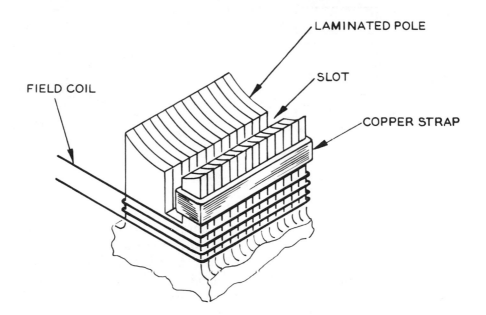

FIELD COIL

LAMINATED POLE

SLOT

COPPER STRAP

Fig. 6-8   Shaded pole

starting coil remains in the circuit after the motor is running.

The shaded-pole motor is amazingly simple and can run indefinitely. It is comparatively inefficient, usually having much less than 50 per cent efficiency. It is used in low power applications such as small fans and phonographs. Speed of rotation is constant, but direction cannot be reversed.

UNIVERSAL MOTOR

In the induction motors discussed earlier in this chapter, the rotors are solid castmetal parts with no electrical connection to the line cord or the field coils. The universal motor works on a totally different principle and is quite different in appearance. Mounted on the motor frame are two field coils wound on laminated pole pieces. The armature is supported between these pole pieces (An arma-

Fig. 6-9 Armature

ture is a rotor with coils wound on it.) The armature is made of laminated metal strips which are slotted, and the armature coils are wound in the slots. Each armature coil is connected to a pair of opposite contacts on a commutator, but are insulated from the core of the rotor. A typical armature is shown in Fig. 6-9 without coils.

The field coils in the frame and one coil in the armature are always in series. For this reason the motor is also called a *series motor.* Electrically, the circuit is as shown in Fig. 6-10. Connections between the armature coil and the field coils are accomplished by brushes contacting the commutator. Thus, for any one position of the armature, one armature coil is directly connected between the opposite sections of the commutator which are under the brushes. The brushes are connected directly to the field coils. As the armature rotates, a new armature coil moves

Fig. 6-10 Circuit of universal motor

into the circuit as each one moves out, so that there is always a complete circuit through the motor.

The commutator, as indicated in Fig. 6-9, is simply a slotted metal ring, with each section insulated from the next by mica strips. Each pair of opposite sections is connected to a separate coil on the armature. The coils are wound in such a manner that the magnetic field in the armature is out of line with that of the field coils. As the armature rotates to line up the two magnetic field, a new armature coil moves into the circuit and the armature magnetic field is again out of line with that of the field coils. As each new successive armature field is pulled into line, the armature continues its rotation.

The brushes making contact with the rotating commutator segments are carbon blocks held in contact with the commutator by a spring. A typical brush is shown in Fig. 6-11. It is a rectangular block curved at one surface to make good contact with the commutator. It is held in a

Fig. 6-11   Brush and holder

rectangular brush holder in which it must fit snugly but should still be free enough to be pushed by the spring to maintain contact with the commutator as the brush wears. The pigtail in Fig. 6-11 is a soft stranded copper wire which connects the brush to the field coil. If a pigtail breaks, the motor might still run since the spring is metal and can carry the current. However, the springs are not designed to carry current, and may get hot and become damaged.

The universal motor has extremely high starting torque and high speed and is used in applications where it is necessary to start a motor under heavy load. For a given horsepower, the universal motor is smaller and lighter than induction motors. The disadvantages of a universal motor are poor regulation, more noise than an induction motor, and radio interference caused by sparking at the brushes. The radio interference can be minimized by placing small capacitors across the sparking circuit, and some manufacturers furnish their appliances with these capacitors already installed.

The speed of rotation of a universal motor can be varied. There are several ways of doing this, all quite different. One method is by means of movable brushes. Recall that the brushes are positioned on the commutator to produce an armature field which is out of line with the field produced in the main field coils. There is one optimum position of the brushes where maximum torque is produced. If the brushes are moved from this position, the motor will still run, but the reduced torque causes reduced speed. It also produces less power. The use of moving brushes is not a popular method of speed control.

A more common method of controlling the speed of a universal motor is tapping the field coils. In effect, the magnetic field of the stator is shortened, producing less torque. This also produces less power as well as reducing the speed. A diagram of a tapped-coil circuit is shown in Fig. 6-12. Only one field coil is tapped. When the switch is at *H* in the figure, the entire tapped coil is in the circuit, and the motor runs at high speed. At *M* and *L,* the coil is progressively shortened, so that the motor runs at medium

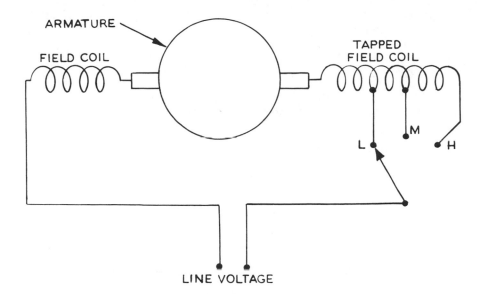

Fig. 6-12   Tapped field speed control

speed and low speed, respectively. A rotary switch is shown in Fig. 6-12, but in practice, push buttons are frequently used on appliances such as blenders. It is not necessary to limit the number of taps to three.

A centrifugal speed control, shown in Fig. 6-13, can be used on any kind of motor. This type of control is sometimes called a *centrifugal governor,* since when it is set it not only determines the speed of operation but maintains that speed under varying loads, even for motors with poor regulation. As shown in Fig. 6-13, there are two contacts which are wired into the circuit so that when these two are touching, or closed, field current must pass through their connection. When they are separated, or open, the circuit is broken, and no current reaches the armature coils. For example, the switch may be in series with one of the brushes. One contact is on the end of an arm which is free to rotate around a pivot point near the other end of the arm. This is the *moving contact.* The other contact of the switch is fixed, although its position is adjustable. When the motor is at rest, a spring pulls the moving contact against the adjustable contact. Now if the motor is turned

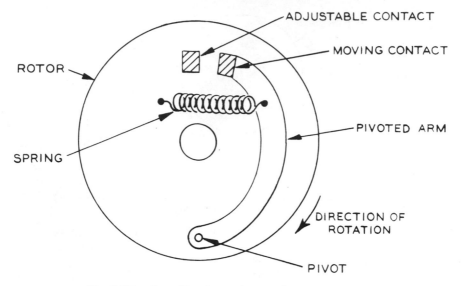

Fig. 6-13   Centrifugal speed control

on, it starts in its normal manner. As it speeds up, a point will be reached where the centrifugal force overcomes the pull of the spring and separates the two contacts, thus breaking the circuit. With no voltage supplied, the motor now slows down. The centrifugal force is reduced, and the spring is able to pull the contacts together again. The motor speeds up, and the process is repeated. This making and breaking of the circuit occurs many times in one revolution of the motor, so that the motor runs smoothly. The speed is set by moving the adjustable contact. When it is moved to the left in Fig. 6-13, centrifugal force will break the contact at a lower speed than when the contact is moved to the right. In effect, the voltage is off a greater portion of the time to produce a lower speed.

In order to prevent pitting of the contacts from the frequent making and breaking of the circuit, it is necessary to include a small capacitor across the contact points. This also minimizes radio interference. The centrifugal switch is more costly than a tapped-coil speed control because of the added parts.

The direction of rotation of a series motor can be

reversed by reversing the armature connections. One way of doing this is shown in Fig. 6-14. Two switches are shown but in practice a single double-pole double-throw switch is usually used. Assume, at any one instant, current enters the field coil at the left and leaves from the field coil at the right, as it does half the time when AC is used. With the switches in the positions indicated by the solid arrows, current flows through the armature coil from left to right. Assume that the magnetic field set up in the armature is then pulled toward the pole pieces by their magnetic fields. Now if the switches are moved to the positions

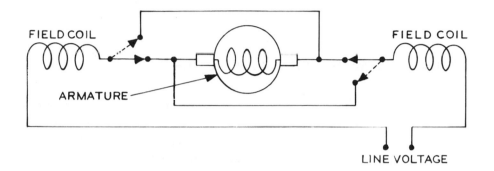

Fig. 6-14   Reversing switch

indicated by the dotted arrows, the current through the armature coil flows from right to left. This produces a magnetic field in the armature which is opposite to the first field, and thus the armature would be repelled where before it was attracted. The armature thus turns in the opposite direction.

Because of its very high starting torque under heavy load, a series motor is used to run a streetcar. In the home it is used for vacuum cleaners, mixers, blenders, sewing machines, power tools, and in other appliances which start and run under *varying* loads. The universal motor is capable of very high speed, and the speed can be controlled by any of the methods described earlier.

SUMMARY OF MOTOR CHARACTERISTICS

The shaded-pole motor is exceedingly simple and rarely requires servicing. It has very low starting torque and runs at constant speed. It may be used for very low-power applications where the load is minimal.

The split-phase induction motor has low starting torque and high starting current. It runs at constant speed. It is used in medium to low power applications in appliances which are not turned on and off frequently. The high starting current causes flickering of lights whenever a split-phase motor is turned on.

The capacitor-start motor has medium torque, medium starting current, and runs at constant speed. It will supply more power than the split-phase motor. It may be used in appliances which are turned off and on frequently, since its medium starting current lessens the annoyance of light flicker. Because of the capacitor, this type of motor is more expensive than a comparable split-phase motor.

The universal or series motor has very high starting torque and can start under heavy load. It has very low starting current. The speed varies and depends on the load. It is used in appliances which may start and run under varying load conditions. For a given size and weight, the universal motor furnishes more power than any of the other motors described above.

Some typical applications of motors are presented in Table 6-1

TABLE 6-1 MOTOR APPLICATIONS

Appliance	Typical rating (horsepower)	Motor type
small fan	less than 1/20	shaded-pole
medium fan	up to 1/3	split-phase
large fan	1/4 1	capacitor-start
mixer	1/10	universal
blender	2/3	universal
vacuum cleaner	1/4 3/4	universal
juicer	1/6	universal
washer	1/4 1/2	split-phase
refrigerator	1/6	capacitor-start
phonograph	1/20	shaded-pole
hair dryer	1/20	shaded-pole
sewing machine	1/4 3/4	universal

REPAIRS

Although there are many different kinds of motors, they suffer similar ailments, and the techniques for locating a source of trouble are much the same for all of them. There are some differences, of course. For example, induction motors do not have brush troubles, and universal motors are not troubled by a shorted starting capacitor. However, most apparent motor troubles are usually not in the motor itself but in the equipment connected to the motor. Although motors are simple to take apart by removing the assembly screws shown in Fig. 6-4, you should not disassemble a motor until you have eliminated all sources of trouble that can be corrected externally.

When an appliance containing a motor fails to operate, first make sure that the outlet has voltage and that the line cord and main switch of the appliance are in working order. The most frequent trouble in vacuum cleaners, for example, is a defect in a line cord caused by constant flexing or tugging. Use the techniques described in Chapter 3 to check the line cord and repair it if necessary. A simple continuity test can be used to check the main switch. If it is defective, replace it with a similar part.

Assuming the line cord and switch are in working order and voltage is getting to the motor, troubleshooting depends on the symptoms first and only secondarily on the type of motor. There are many different kinds of symptoms:

(1) Nothing happens when a motor is turned on. It won't start and is silent.

(2) The motor won't start, but it hums or tries to start.

(3) The motor runs, but gets hot.

(4) The motor is noisy.

(5) In a universal motor, the brushes spark.

(6) The motor runs but has no power.

NOTHING HAPPENS (1)

If a motor won't start and is silent, even though voltage is reaching it, the trouble is probably in the motor. However, many motors have a *thermal overload protector,* which is a switch that shuts the motor off when it gets too hot. Before the motor can be started again, this switch must be closed manually, which is done by pushing a *reset* button on the outside of the motor case. Look for such a button (it's usually red) and push it before making any other test. If there is no thermal overload protector, or if the motor still won't start after pushing the reset button, it is necessary to consider the type of motor. In any motor, a burnt-out winding will prevent starting.

In a split-phase motor, the trouble could be an open field coil, an open starting coil, or a centrifugal switch stuck in the open position. Look again at Fig. 6-3, repeated here for convenience. Note that a

RUNNING COIL STARTING COIL RUNNING COIL STARTING COIL

ROTOR ROTOR

CENTRIFUGAL SWITCH CENTRIFUGAL SWITCH

115V **(a) START** 115V **(b) RUNNING**

Fig. 6-3 Repeated for convenience

continuity check across the line (with the power disconnected, of course) would not tell you anything, since
there would be continuity even if *one* of these three
elements were open. However, if the running coil is good,
the motor should run once it is started. Plug in the motor
and spin the rotor with your fingers. If the running coil is
good, the motor will run, and therefore the trouble is in the
centrifugal switch or the starting coil. If the motor still does
not run, the trouble is probably in the running coil (field
coil). With the power disconnected, hold the centrifugal
switch open with your fingers or piece of cardboard and
make a continuity check across the field coil. You will
probably find it open. It is possible to buy a new field coil
assembly to replace a defective one, but if the motor is old,
it pays to replace the whole motor if a coil is open. In rare
cases, you may be able to spot the break and solder the ends
together, but it is a difficult job for an amateur, and a motor repairman would charge more than a new motor costs.

If the motor runs after you spin it by hand, check the
centrifugal switch. With the power disconnected, you
should be able to open and close this switch easily with

your fingers. If the contacts are dirty or pitted, they may not be making an electrical connection. They can be cleaned with sandpaper. Note: *never use emery paper,* since the grit is conducting and can cause a short circuit if any particles fall into the motor. A quick check on a centrifugal switch can be made by shorting it with a clip lead. If the trouble is caused by an open switch, the motor will start with the clip lead removing the switch from the circuit. If the centrifugal switch is stuck in the open position, clean the moving parts and add a drop of oil on its pivot point. Be careful not to use too much oil, since excess oil may leak between the contacts of the switch and form an insulating coating which prevents the switch from closing.

If the centrifugal switch is good, but the motor will not run unless you spin it with your fingers, the trouble is probably an open starting coil. You can make a continuity check across it, holding the centrifugal switch open with your fingers or a piece of cardboard. If the starting coil is defective, it doesn't pay to fix it. It can be replaced, or if the motor is old, it is better to replace the whole motor. In general, it doesn't pay to try to fix burnt-out or open coils.

In a capacitor-start motor that won't start, the trouble could be any of those areas described above for the split-phase motor, and in addition it could be caused by a defective capacitor. Try spinning the motor by hand to start it. If it now runs, the trouble could be in the capacitor, the centrifugal switch, or either of the windings. Look again at Fig. 6-6. The capacitor should look like an open circuit or a very high resistance. Check the two coils for continuity (as described above for the split-phase motor) and make sure the centrifugal switch is operating properly. As before, you should be able to move it easily with your fingers, and the contacts should be clean. If the capacitor is open, it cannot be checked by a continuity test since it would show no continuity when it is good as well as when it is open. Disconnect the capacitor and put it across a voltage source for a few seconds. If the capacitor is good, when you remove it from the voltage source, there will be a voltage stored in the capacitor. Now bring the leads of the capacitor together, and they should produce a spark when they touch. Use insulated pliers to hold the capacitor leads when you do this. If there is no spark, the capacitor is probably defective and should be replaced. Since capacitors are a frequent source of trouble, mail order houses carry replacement capacitors for many appliances.

If a universal motor won't start, the trouble is usually in the brushes. By referring to Fig. 6-10, you can see that a continuity test can be used to check a universal motor. Occasionally the trouble will be a worn out brush. Check to make sure brushes are being held in contact with the armature. If they are worn, replace them. Note that one end of the brush is curved to approximately the shape of the commutator, as shown in Fig. 6-15. When new brushes are installed, they will spark for a few minutes until they are worn to an exact fit on the commutator. Don't worry about this sparking for the first five minutes, but if it persists much longer, make sure brushes are not too loose in their holders. If the curved part of the brush does not mate with the commutator, you can grind it in as shown in Fig. 6-16. Wrap a piece of sandpaper around the commutator with the grit side out. Then put the brushes in the holders and turn the commutator by hand. The sandpaper will wear away the brush until there is a good fit. The brushes in motors in most appliances can be removed from the outside without taking the motor apart.

PIGTAIL

Fig. 6-15   Brush

If a field coil is open, you should replace the whole stator. These open coils can be located by the usual continuity checks. In the armature, make continuity checks across each pair of opposite segments of the commutator. Each pair is attached to a different coil, and if one of them is open, the whole armature should be replaced. Also check continuity between each commutator segment and the body of the armature. You should read an open circuit here, since continuity would indicate a coil short-

RUNNING COIL   STARTING COIL
CAPACITOR

ROTOR

CENTRIFUGAL SWITCH

115V   (a) START

RUNNING COIL   STARTING COIL
CAPACITOR

ROTOR

CENTRIFUGAL SWITCH

115V   (b) RUNNING

Fig. 6-6 Repeated for convenience

ing to the body. If an armature has a shorted coil, replace the whole armature. Don't try to rewind the coil. It's not worth the effort, since a new armature is relatively inexpensive.

THE MOTOR HUMS, BUT WON'T START (2)

In any kind of motor, tight bearings, a bound load, or other mechanical trouble can prevent the motor from starting, although it will hum and try to start when it is turned on. Note: *disconnect it quickly,* since a motor can burn out under these conditions. A simple check is to try to turn the motor by hand *with the power off.* It should turn easily. If the rotor cannot be moved easily, there are three possible sources of trouble. First check the load. To do this remove any belts or gears connecting the motor to the load. In a blender, remove the glass bowl. Try to turn the motor again. If it now rotates easily with the load detached, the fault is not in the motor, but in the load. Look for mechanical troubles, such as bent linkages, tight or broken bearings, or rust or dirt. Clean and repair the load as required.

If the motor is still bound with the load detached, the trouble is caused either by the rotor rubbing against the stator or by a tight or frozen bearing. When a bearing is tight or frozen, it may simply need cleaning or oiling. Be careful not to use too much oil since an excess can spatter, and if oil drops get on the contacts of a centrifugal switch or between brush and commutator, they can break the electrical circuit. If a bearing cannot be cleaned or remains bound, replace it. When cleaning or replacing a bearing make sure you observe exactly how the washers and shims are positioned on the end of the spindle, and when you reassemble the motor, put them back the same way. These shims prevent too much endplay in the armature so that the brushes are positioned correctly on the commutator.

If the rotor rubs against the stator, a broken bearing may be the culprit. Also, an end bell may be bent or broken. In either case replace the defective part with one that is identical. Motor parts are available from authorized service dealers or from the manufacturer at the address on the nameplate.

If an induction motor has a free-spinning rotor but still hums without starting, the trouble can be in the centrifugal switch. The switch may not be closing perfectly because of dirty contacts. Clean the switch contacts with sandpaper. Make sure the parts of the switch can be moved freely. If the rotor has too much endplay, the vibration can prevent the switch from acting properly. Endplay should be at least 0.005 inch and may be up to five or six times this amount. Note that an open centrifugal switch was also given as a cause of failure if the motor didn't start and was silent. The difference here is that the points are in close enough contact to move the rotor but the current separates the points before the rotor moves. In this case, as before, you can start the motor by spinning it, and then it will run.

In a capacitor-start motor, the motor can still run if

BRUSH

SANDPAPER

COMMUTATOR

Fig. 6-16 Fitting brush to commutator

the capacitor is shorted. This is shown in Fig. 6-17. In (a), you see the electrical circuit of a split-phase motor, and in (b) that of a capacitor-start motor. Note that if the capacitor is shorted in (b), the circuit becomes identical with that of the split-phase circuit in (a). Recall, however, that the capacitor was added to give higher starting torque, and the capacitor-start motor is usually used in applications where this extra starting torque is needed. Thus, when the capacitor is shorted, the motor no longer has high starting torque. Since this type of motor is usually connected to a heavy load, the shorted capacitor prevents it from turning the load, although it strains to do so, and hums while trying. The solution obviously is to replace the capacitor.

THE MOTOR OVERHEATS (3)

Tight bearings or a heavy load can cause a motor to overheat. When the motor has cooled, disconnect it and try spinning the rotor with your fingers. If it fails to turn easily, look for bearing trouble, rotor rubbing against stator, or load trouble. These faults and their remedies are discussed in the two preceding sections.

If the rotor spins easily, but the motor gets hot, a possible trouble is a centrifugal switch which remains closed. Referring to Fig. 6-3 you can see that when the motor is in the start position, two coils are across the line. After the motor is running, the centrifugal switch is supposed to open, as in Fig. 6-3(b), and only one coil remains. If the switch fails to open, the starting coil remains in the circuit and continues to draw current. This excess current causes overheating. To remedy this make sure the centrifugal switch can be moved easily with your fingers. If the contacts are burnt shut, separate them and clean them with sandpaper. *Never use emery paper,* since the dust from emery paper conducts electricity and can cause a short circuit if it falls in the motor. Make sure the parts of the switch move easily. If necessary, oil the pivot lightly. Warning: too much oil is bad since the excess can leak between the contacts and prevent them from closing.

If a motor gets too dirty, it will also tend to overheat. Many motors have a built-in fan for cooling, but if the air intake slots are clogged, the fan will labor in vain. Blow out or vacuum the motor and make sure all air vents are open. An old toothbrush can be used to clean accumulated dirt off the stator. Never use any sort of cleaning liquid or water on the coils or internal parts of the motor.

Anything that causes a motor to draw excessive current is bad, since excessive current causes overheating. If stator windings become shorted or grounded to the case, the motor will run but will draw excessive current and overheat. If you have an ammeter, you can measure the current and compare it to the rating on the nameplate. Without an ammeter you can still get a good idea of the current being drawn by observing the electric meter which registers the amount of electricity used. This meter is located outside your house. A typical meter is shown in Fig. 6-18. In the center of the face there is a slot, and a horizontally mounted rotating disc protrudes through this slot. Whenever electric current is being drawn anywhere in the house, this disc rotates, and the speed of rotation is directly proportional to the current. Make sure that all electricity in the house is off. Disconnect all appliances such as refrigerators or furnaces which may start of their own accord. You will then note that the disc on the meter is stationary. Now turn on something with a known current

RUNNING COIL   STARTING COIL

(a) SPLIT PHASE

RUNNING COIL   STARTING COIL

CAPACITOR

CENTRIFUGAL
SWITCH
LINE

(b) CAPACITOR START

CENTRIFUGAL
SWITCH
LINE

Fig. 6-17   Comparison of induction motors

or power rating. For example, turn on a 200-watt lamp. You will see that the disc is now rotating. There is a black smudge at one point of the disc's circumference, and by counting the number of times this smudge passes you can count the revolutions of the disc. Assume the disc makes two revolutions per minute when the 200-watt lamp is on. Now if you turn off the lamp and turn on a 1000-watt toaster, you would be using five times as much current and the disc would turn five times as fast or ten revoltuions per minute. Since, you know that power in watts is volts times amperes, you can calculate that a 200-watt lamp will draw slightly less than two amperes from a 115-volt line. Suppose a motor is rated to draw six amperes. You would expect it to make the disc rotate about three times as fast as the lamp or about 6 revolutions per minute. Actually, if it rotates seven or eight times per minute it would not be excessive, but ten or twelve revolutions per minute would indicate that the motor is drawing excessive current. If you find that the motor is drawing too much current, check the coils for shorts or grounds. If coils are bad, they should be replaced.

THE MOTOR IS NOISY (4)

Noise in a motor is usually a mechanical problem. Most likely something is loose. Go over the motor carefully and tighten all housings, baffles, and other attachments. Noise can also be caused by excessive endplay of the rotor or by a loose bearing. Tighten bearings if needed. Possibly a shim or washer was omitted in assembly. Check the endplay and add small thrust washers to reduce the endplay to something between 0.005 and 0.025 inch. Noise may also be caused by loose parts in the load, such as a loose belt, pulley, gear, or fan. This is simple to track down, and corrective measures are obvious.

Dirt inside a motor can also cause noise, especially if the dirt gets between the stator and rotor. Dirt includes floor, as well as dust and lint. In mixers especially, all kinds of unwanted materials accumulate in the motor. To clean a motor, first take it apart and blow off all the loose dirt or use a vacuum cleaner to pick it up. Make sure the vacuum cleaner doesn't pick up loose screws, washers, or other parts of the motor. Use an old tooth brush or a wooden tool to clean dirt from the windings. Never use liquid on any part of the motor if the liquid can get to the windings. Bearings, shafts, and the centrifugal switch can be cleaned in solvent if they are first detached from the motor. Make sure they are dry before reassembling the motor. Put a drop of oil on each bearing before reassembling.

THE MOTOR IS SPARKING AT BRUSHES (5)

Excessive sparking at the brushes of a universal motor is a warning signal of trouble. If there is not something already wrong with the motor, there soon will be. Fortunately, the trouble is usually in the brushes themselves and is easily remedied. Remove the brushes and check the way they fit against the commutator. If the fit is bad, put sandpaper around the commutator, as shown in Fig. 6-16 and discussed above, and rotate the armature until the brushes are sanded to shape. If a brush does not slide easily in the brush-holder, there will also be sparking, due to poor contact between the brush and the commutator. Also, if the brush is worn down so that the spring no longer pushes it against the commutator, or if the spring itself has lost its tension, there will be sparking. Put in new brushes or springs or both, as required. New brushes must fit snugly in the brush-holders but should slide easily. If they bind, you can sand down the rough spots.

A more serious cause of sparking at the brushes is a shorted or open coil on the armature. If the brushes seem to be in good condition, check continuity between opposite segments of the commutator and between each segment and the body of the rotor. If the armature has a defective coil, replace the armature. *Do not* try to fix the coil.

The commutator itself can be damaged by poor brushes and can then cause trouble after the bad brushes are replaced. When a motor runs properly, there will be a shiny ring around the commutator where the brushes make contact. This is shown in Fig. 6-19. If one segment of the commutator protrudes from the surface, it will push up the brush which then bounces on the next segment and causes sparks. These sparks burn pits in the commutator and later when the high segment is fixed, the arc burns can cause more arcing. Such a pitted segment is shown in Fig. 6-20. Similarly, if there is arcing caused by a bad brush, there will be pitting all around the commutator. After the brush is replaced, there may be arcing caused by the pits. Whenever new brushes are added, the commutator should be cleaned and sanded if necessary.

When a commutator has seen many years of service the shiny ring may become a groove around the commutator, as shown in Fig. 6-21. This is not necessarily bad. However, if the surface falls below the level of the mica insulators, the brushes may bounce on the mica. Use a hacksaw blade to cut off any mica that protrudes above the surface of the commutator where the brushes might strike. Also, whenever a commutator is sanded to remove pit marks, make sure the mica insulators do not project

KILOWATT HOURS

ROTATING DISC

Fig. 6-18   Electric meter

above the surface. Use a hacksaw blade to trim off excess mica.

THE MOTOR TURNS BUT HAS NO POWER (6)

If a motor runs but lacks power, first look for mechanical troubles. With power off make sure the motor and its load are free to move without restriction. Bearings which are too tight or need oil can hamper operation. Also, if the endplay is less than the minimum 0.005 inch, the expansion due to heat when the motor is running can cause the rotor to bind. If the motor does not turn easily, look for a mechanical defect in bearings, belt, gears, fan, and other likely places and loosen or replace parts until the motor turns easily.

If the capacitor is shorted in a capacitor-start motor, the motor will lack power on starting, but once it is running, the centrifugal switch will remove the capacitor from the circuit, and the motor should run satisfactorily. On any

Fig. 6-19   Armature with good commutator

Fig. 6-20   Pitted commutator

induction motor, if the centrifugal switch fails to open, the motor will be hot and inefficient and will draw too much current.

As a final check in any motor, make sure there is no voltage on the frame of the motor while the motor is running. You can do this by putting one probe of your meter or test-light on the frame and the other on ground. Voltage on the case is a shock hazard, but in addition it may be a sign of a coil shorted to the case, which would eventually cause failure.

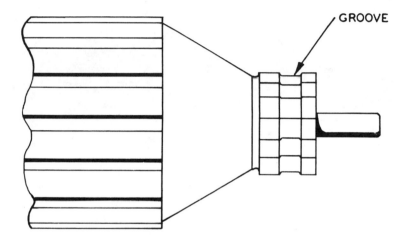

GROOVE

Fig. 6-21   Grooved commutator

Chapter 7

Control Devices

Automatic appliances such as dishwashers and clothes-washers perform a number of different functions, such as washing, rinsing, and drying. These functions must be performed in the correct order. Thus, in addition to the switches which start and stop each of the functions, there must be some sort of "brain" which determines when each event shall occur. The switches are control devices in that they control individual functions. The "brain" is also a control device since it controls the switches.

In a sense every electrical appliance needs at least one control device, even those with no switches. For example, if a popcorn popper starts to heat up when you plug the cord in an outlet, the control is plugging it in. Most appliances, however, do have switches to start them, and these switches take a variety of shapes and forms. In appliances which are connected to plumbing there are also mechanical switches to start and stop the flow of water, but these mechanical switches are usually controlled electrically.

In an automatic appliance, the master control determines which switches turn on and off. The determination may be based on time or on a physical condition, such as water reaching a suitable level or an oven reaching a desired temperature. Temperature sensing switches, or thermostats, are considered in detail in Chapter 4. They are used in most heating appliances. Other control devices, (switches, timers, and solenoids) are discussed in this chapter.

SWITCHES

Consider the many different ways you "turn on" electricity. The common wall switch for lighting is flicked up or down. Some blenders have a set of pushbuttons for selecting the proper speed. Pushbuttons are also used on some heaters to select the proper temperature. The tip-over switch on a heater is a pushbutton type switch which must be held in to keep the electricity connected. Another type of pushbutton switch is found on many electric juicers. It is pushed and released to turn the appliance on, and again pushed and released to turn it off. A tank-type vacuum cleaner usually has a switch which is operated by stepping on it. This is essentially a foot-operated pushbutton switch. Some appliances have a rotary switch, as for example, a dishwasher or clotheswasher. Some mixers have a sliding lever, where the amount of movement regulates the speed of the mixer. A toaster is turned on by depressing the lever which lowers the bread.

Regardless of the external form of the switch, they all have one thing in common. A switch simply connects two contacts when it is in the *on* position and separates the contacts when it is *off*. If you checked continuity across a switch, it would look like an open circuit when it is off and a short circuit when it is on. This is shown in Fig. 7-1. In (a), the switch is open, and thus when the probes from

Fig. 7-1   Continuity across switch

the ohmmeter are placed across it, the meter indicates infinite resistance (∞) which means an open circuit. In (b), with the switch closed, the meter indicates zero ohms, or a short circuit. When checking the continuity of a switch, you must make sure that there is no other continuous path across the switch in the rest of the circuit.

A switch can fail by sticking in the open or closed position. If this happens, the switch should be replaced. Disconnect all wires to the switch, noting carefully where each wire is connected, and put in a new switch. Connect the wires to the new switch in the same manner as they were connected to the old switch. Sometimes one switch in an assembly of several pushbutton switches goes bad, and it may be necessary to replace the whole assembly because the switches cannot be separated. Here, many wires have to be disconnected and reconnected, and you should make a drawing of the connections before unsoldering a single wire.

A switch that simply makes and breaks one connection, like that in Fig. 7-1, is called a *single-pole single-throw switch.* "Single-pole" refers to the fact that only one wire is connected or disconnected by the switch. "Single-throw" means that there is only one method of connecting this wire. In more complicated switches, one or more wires may be connected at different times to one

Fig. 7-2   Single-pole, double-throw switch

or more contacts. For example, in a rotary switch, the wire connected to the rotating contact may be connected to any of several contacts in turn as it is rotated. A *single-pole double-throw* switch, which might be used to select a high or low heat in a heating appliance, is shown in Fig. 7-2. Note that with the switch in the center position shown, the appliance is off. If the switch is moved to the *H* contact only the high heating element is in the circuit. In the *L* position only the low element is connected. There are two *on* positions, hence "double-throw". Externally the single-throw and double-throw switches may appear identical, but the double-throw has three contacts whereas the single-throw has two.

When checking a double-throw switch for continuity, you must take into careful consideration what you are doing and what the rest of the circuit is. For example, in Fig. 7-2, if you measured the continuity between the *H* and *L* contacts, you would always read the sum of the two heating resistances, regardless of the position of the switch. To check this switch it should be disconnected from the rest of the circuit first. Now there should be an open circuit between two of the contacts regardless of the position of the switch. These would be the "throw" contacts, shown as *H* and *L* in the figure. Between the third or *common* contact and either of the others there should be

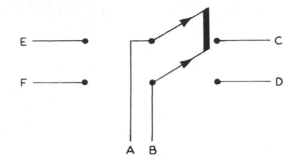

Fig. 7-3   Double pole, double-throw switch

an open circuit in one position of the switch and a short in the other. If the switch is bad and needs replacing, make sure that the common wire is connected to the common contact of the new switch.

A *double-pole double-throw* switch is shown schematically in Fig. 7-3. When the switch is closed to the right, wire A is connected to wire C, and wire B to wire D. When the switch is moved to the left, A is connected to E, and B to F. In effect, there are two switches which are tied together physically so that both are thrown simultaneously by moving only one lever. An application of the double-pole, double-throw switch is shown in Fig. 6-14, as a reversing switch for a series motor. In this application the switch is thrown to one pair of connections or the other. In other applications, there may be a *neutral* or *off* position between the two, but the switch is still double-throw.

A common application of a single-pole, double-throw switch is to control a ceiling light fixture from two switches at either end of a room. Two such switches are used, as shown in Fig. 7-4. When both switches are connected to the same wire running between them, the circuit is closed and the lights light. If either switch is thrown to the other wire, there is no continuous circuit and the lights go out. Note that from the off position, the lights will go on if *either* switch is moved.

LIGHTING FIXTURE

115 V

Fig. 7-4   Two controls for lighting

An unusual application of switching is in an electric range. Using two identical Calrod elements in one heating unit, with appropriate switches, it is possible to get five different degrees of heat. This is indicated in Fig. 7-5 where two 80-ohm elements are shown. The range is serviced by a three-wire line, where one line is at ground and each of the other two are 120 volts from ground. The phases of the two hot lines are such that they differ by 240 volts.

In (a), both 80-ohm elements are in parallel across 240 volts. Each draws three amperes, and the total power in both is 1440 watts. In (b), the 240-volt line is across one 80-ohm element to produce 720 watts. In (c), each of the elements is across 120 volts and draws 1.5 amperes. The total current of three amperes produces 360 watts. In (d), only one element is used across 120 volts, producing 180 watts. Finally in (e), both elements are in series across the 120-volt line. The 160-ohm resistance draws 0.75 ampere, and the power is only 90 watts.

THREE-WAY LAMP

A very common appliance, found in most homes, is a three-way lamp. In one version, two or more bulbs are switched on separately or in combination. In another, a single three-way bulb is used which has two filaments.

Fig. 7-5   Range connections

Either of these filaments can be switched on separately or both together. Two popular types of three-way bulbs are the 100-200-300 watt and the 50-200-250 watt. In both the multiple bulb version and the three-way bulb, the degree of lighting is controlled by a switch. This is usually a rotary switch with a cam-like blade, as shown in Fig. 7-6. The three-way bulb itself has two filaments. In the 100-200-300 watt bulb, the low filament draws 100 watts and the high filament draws 200 watts. In the 50-200-250 watt bulb, the low and high filaments draw 50 and 200 watts, respectively. In the figure the two contacts are numbered 1 and 2. With the switch blade as shown in Fig. 7-6, neither filament is connected, and the lamp is off. As the switch is rotated, detents stop it at the proper contact positions. In the first position, the narrow part of the

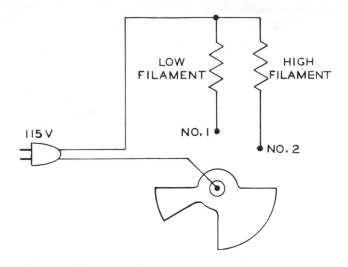

Fig. 7-6 Three-way lamp

blade contacts point 1 and closes the circuit to the low filament. In the second position, the narrow blade touches point 2, and the opening between the blades falls in the vicinity of point 1. Thus, only the high filament is connected. In the third position, the wide blade touches both points 1 and 2, and both filaments light. Finally, the next position is a return to the off position shown. In the lamp with separate bulbs, the switch operates in the same manner.

A three-way lamp can develop trouble in the switch itself or in the contacts to the bulb. If the switch is noisy, it indicates arcing inside which is usually a sign that the switch will soon fail. As with most switches, it doesn't pay to try to repair it. Simply replace the switch with a new one that has the same rating. In some lamps the switch can be separated from the socket; in others the two are integral. Before buying a new switch, open the housing and see just what it is you need.

To take apart the switch housing, look for small screws which hold it together. Some housings are held together by friction and can be pulled apart. Notice in Fig. 7-6 that one wire from the power line goes directly to the bulb filaments and does not go through the switch.

However, in the lamp, you will find all the wires in a bunch and you must separate them to determine which is the *common* wire. You usually will leave the common wire undisturbed and remove only the wires actually connected to the switch. Note how they are screwed on to the switch terminals and replace them in the same manner on the new switch.

TIMERS

Large automatic electrical appliances must be controlled so that each function is performed in the proper sequence. In a dishwasher, for instance, first a pump sucks out any water which has accumulated, then a valve opens to let in hot water. When the water level is correct, the valve closes, and the motor turns on and sprays the soap and water over the dishes. The motor subsequently stops, and the water is pumped out. Clean water enters and the process is repeated several times. Finally a heating element is activated to dry the dishes. Each function is turned on and off by an electric current, but a master control is needed to open and close the switches at the proper time in the cycle. The master control is called a *timer.*

The heart of the timer is simply an electric clock motor which runs at a constant speed. A set of gears reduces the speed so that the motor rotates a spindle at the rate of one revolution for one complete operation of the appliance. This is similar to the gear train in an electric clock which steps down the motor speed so that the minute hand makes one revolution per hour. The exact gear ratio varies with the appliance. If the complete wash-rinse-dry cycle should take 55 minutes, for example, the gear ratio will allow the spindle to make one revolution in 55 minutes.

Mounted on the spindle and firmly attached to it are several cams, as shown in Fig. 7-7. These cams have projections on them to actuate switches, and each cam controls a separate function of the appliance. Thus, one turns the motor on and off, another controls the valve for water

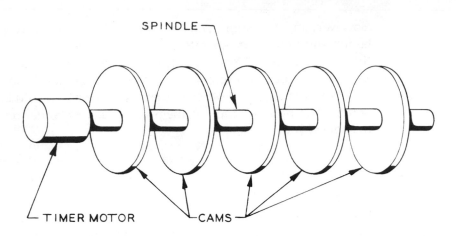

SPINDLE

TIMER MOTOR   CAMS

Fig. 7-7   Timer and cams

flow, etc. A typical cam and the switch it controls are shown in Fig. 7-8. A spring rides on the edge of the cam and when it is raised by a projection, it closes a pair of contacts, as shown in (a). In (b), the spring is between projections and the contacts are open. The number of projections on each cam and the length of each projection correspond to the number of times the switch must be closed and the length of time it remains closed.

If the cam turned at a constant speed, the switches would open and close slowly, and the current flowing at the moment of opening would cause sparking which would pit the contacts, possibly welding them shut or preventing them from making good contact. To avoid this difficulty the spindle is spring-loaded in such a manner that it moves in short impulses instead of turning smoothly. Each impulse is rapid, but with the pauses between, the total time of rotation is still held to one complete cycle of the appliance.

This type of timer consisting of cams and switches is the most common type, but there are others. A somewhat different kind of timer has a long screw in place of the spindle. A carriage is threaded on the screw and moves horizontally from one end to the other as the screw rotates. Metal fingers on the carriage make appropriate contacts, closing circuits, as it moves along.

Timers   **146**

FIXED CONTACT

FIXED CONTACT

SPRING

MOVING CONTACT

SPINDLE

(a) CLOSED

(b) OPEN

Fig. 7-8   Cam and switch

A timer motor rarely needs servicing, but if it fails, it should be replaced by an identical part. Before removing the old timer motor, however, make sure that it is receiving voltage at its terminals. If there is voltage, but the timer motor does not run, see if there are mechanical restrictions, such as stuck bearings in the gear box, before deciding the motor is bad.

A more common fault in a timer is a pair of switch contacts which are welded shut from arcing. The switch levers should move easily when pushed with the fingers and should spring open when released. If contacts are stuck together, pry them apart, being careful not to bend the springs, and clean them with sandpaper. Contacts also may be pitted and burned so badly that they fail to make an electrical connection. Again, they should be sandpapered smooth. Never use emery paper, since the grit is conductive and can cause a short circuit if particles fall in the wrong place.

If a cam gets loose on the motor spindle, you can tighten it, but first you must find out where in the cycle the particular switch closes. Some manufacturers supply this information in the instruction sheets in the form of a timetable for the different functions. You can also get the information from an authorized service dealer. If the cam is twisted on the spindle, the switch will open and close at the wrong times.

SOLENOIDS

A *solenoid* is an electromagnet. Basically it consists of a coil of wire and an iron bar which is free to move through the coil. A spring holds the bar outside of the coil when the circuit is disconnected, as shown in Fig. 7-9. When the coil is energized, that is, when a voltage is applied, current flows and makes the coil a magnet. The iron bar is pulled into the center of the coil. When the voltage is disconnected, the spring pulls the bar out again. The bar can be attached to a mechanical linkage to shift gears, tighten a drive belt, open or close a valve, start or stop an agitator in a washer, or perform other mechanical functions.

An example of a water valve, using a solenoid, is shown in Fig. 7-10. When the coil is energized, the pointed iron rod is pulled up, as indicated, and water flows through the pipe to the tub. When the electricity is turned off, the bar is pushed down and blocks the pipe, stopping the flow of water. The coil itself is usually a separate entity

Fig. 7-9   Solenoid

115V

WATER INLET

TO TUB

Fig. 7-10 Water valve

which is slid over the outside of the valve and is easily removable for testing or replacing.

A *relay* is a solenoid which controls a high current or voltage. For example, some of the functions in a dishwasher, such as drying, draw currents which are too great for the small contacts in the timer. The small contacts close a circuit to energize a solenoid, and the solenoid closes a pair of heavier contacts to start a larger current. The basic relay circuit is shown in Fig. 7-11. The solenoid coil is wound permanently on an iron core to make a strong electromagnet.

When the switch is closed, current flows through the coil making it a magnet. It attracts the movable soft iron bar holding one heavy contact. Since one end of this bar is fixed on a pivot point, the bar rotates on this pivot, and in doing so, the heavy contact on the bar is brought against a *fixed* contact closing the electric circuit to the load. The voltage source can be a heavy-duty 240-volt line or a battery or whatever is required. When the switch is

Solenoids   **149**

Fig. 7-11 Relay

opened, the solenoid no longer attracts the movable arm, and the spring pulls the arm away and thus separates the contacts. The arm is shown in the open position in Fig. 7-11.

The most common trouble in a solenoid or relay is an open coil or a grounded coil. If a solenoid fails to operate, make a continuity test at its terminals. You should read a small resistance on an ohmmeter, the resistance of the coil. If you use a test lamp, the light should glow, indicating continuity. Also check continuity between each of the solenoid terminals and the case. Here you should have no continuity unless the coil is grounded. If the coil is open or grounded, it can usually be replaced easily. In some small relays, it may be cheaper to replace the whole relay.

Check relay contacts for continuity when they are closed and if necessary, sandpaper them smooth. Many troubles with relays and solenoids are mechanical. The pivot point in the relay may need a drop of oil. Valves which are operated by solenoids may be stuck. In general, if the solenoid coil is all right, the trouble is not in the solenoid, but is probably a mechanical problem in the part that is operated by the solenoid.

Chapter **8**

Small Appliances With Motors

Many appliances use some sort of motor to move a tool to do a particular job. For example, in a fan a motor turns the fan blade; in a mixer, the motor rotates the beaters; in a blender, a knife is turned; and in an electric can opener, the cutting wheel rotates. Power tools, dentist drills, and even some toys use a motor for nothing more than rotating or moving a tool, just as in home appliances. In this chapter, some of the home appliances with motors are considered, although the principles of operation and methods of repair can be applied to shop tools and professional instruments as well.

ELECTRIC FANS

An electric fan is a very simple appliance consisting simply of a switch and a motor which turns a fan blade. There are two types, one used for moving air for cooling and exhaust fans to remove kitchen odors, but they are both the same electrically. Fans almost always have an induction motor, as indicated in Table 6-1 on page 121. Most fans used in the home are less than 1/4 horsepower and use either a shaded-pole or a split-phase motor. Both of these have low starting torque, but air is not a very heavy load. Some fans have a universal motor for use in areas where only direct current is available, but these are rare. Thus, for most fans, brush troubles are nonexistent.

Kitchen exhaust fans are usually controlled by a switch near the stove. This switch connects the outlet to the power line. A sketch of an exhaust fan is shown in Fig. 8-1. The fan is usually mounted in the ceiling or high on the wall above the kitchen range. The fan motor is mounted on two to four metal arms which are attached to the studs by rubber mountings to reduce noise from vibration. A short line cord from the motor is plugged into a single outlet which is controlled by a switch near the stove, as mentioned above. Since this cord is never moved, you never have to worry about line cord troubles in an exhaust fan. At ceiling level, an ornamental grill of metal or plastic hides the fan but allows the hot air and *grease* to get through. Above the fan there is some sort of covering (not shown in the figure), which remains closed when the fan is not in use, but opens when it is turning. This keeps out rain and cold. Exhaust fans run at one constant speed.

The motors in most exhaust fans are "permanently" lubricated. This means that they have large wicks soaked in oil in each end bell which should keep the bearings oiled for many years. Nevertheless, after many years of operation, these oil wicks do get dry and the motor needs reoiling. If the oil wicks are dry, one symptom is that the fan turns too slowly. To check this, remove the grill and spin the blades with your fingers. The blades should coast

Fig. 8-1   Exhaust fan

easily and slowly come to a stop. If the fan stops abruptly or seems to have friction in the bearings, it probably needs oiling. Look for oil holes, and if you are lucky enough to find them, you can oil the fan without detaching it from its mounting. Use light oil and run the motor a few times between successive oilings. If there are no oil holes, you must remove the motor and take off the end bells. Add oil to the cloth wicks until they are saturated, and reassemble the motor. Wipe off any excess oil that has leaked outside the motor.

If an exhaust fan won't run at all, check the motor for continuity. You can do this simply by touching your meter probes to the prongs of the line cord with the switch in the on position. If the circuit is open, the trouble may be in the switch or the field winding of the motor. Usually, however, an exhaust motor fails to run because it is clogged with dirt. Dirt in this case means flour, grease particles, and everything else that is sucked up through the grill when the fan is running. Clean the grill and all parts of the motor with paper towels. If grease has

Electric Fans   153 gotten inside the motor, take the motor apart and clean it. A good cleaning solution is simply detergent and water. You can use this on the grill, the end bells, the rotor, and bearings, but don't use fluid on the stator windings. If the contraption above the motor is covered with grease so that it fails to open, this will also need cleaning.

After you have cleaned the motor, reassemble it. Make sure you oil the bearings before you put it back on its mounting. It is a good idea to plan to clean a kitchen exhaust fan regularly before waiting for it to develop troubles. Twice a year should be sufficient.

Fans for cooling a room by moving air come in a variety of sizes and shapes. They are usually mounted in a round or square guard frame to prevent people from sticking their fingers into the whirring blades. Incidentally, if you are holding the blade of a small fan when it is turned on, the blade will not turn because the motor has low starting torque. Don't try this, however, on large fans which use capacitor-start motors. Cooling fans usually have some speed control arrangement.

Most fan troubles are mechanical rather than electrical. The fan is dropped or knocked over and a blade is bent or rubs against the frame. If a fan vibrates or is noisy, a blade may be out of balance from being bent. Rotate the fan by hand and make sure each blade is the same distance from some fixed reference point on the frame. A simple way to do this is to hold a pencil against a bar of the frame so that it just touches a blade, as shown in Fig. 8-2. Then hold the pencil steady and rotate the blades. Each blade should just touch the pencil. If necessary, bend the blades carefully until they do so.

If the fan blades hit against the frame, you can bend the frame away so that they turn freely. You should then check as above to make sure the blades are balanced.

If an oscillating fan fails to oscillate, but runs smoothly otherwise, the problem is caused by stripped gears or a worn clutch in the oscillating linkage. If you don't need the oscillating feature, you may decide to do nothing, since the fan and motor still operate properly. However, if you wish to repair this, you should replace the whole

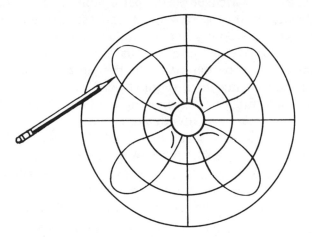

Fig. 8-2   Checking blades

linkage, since if one part is worn, others will soon wear out. When a new linkage is installed, put a drop of light oil on each pivot point and bearing.

If any type of fan won't run at all, check to see if there is voltage at the outlet first. Then your next check should be a continuity check across the prongs of the line plug, with the switch turned to the on position. Note that this checks line cord, switch, and motor windings at one time. If an open circuit is indicated, then the line cord, switch, and motor should be checked separately. For line cord tests and repairs consult Chapter 3, and for motor maintenance consult Chapter 6. Dirt clogging the fan blades or motor can also stop a fan. Dirt can be washed off blades and mechanical parts using water and a detergent, but the motor windings must be kept dry. Use a wooden stick or an old toothbrush to remove dirt from the motor.

If it is necessary to take apart the motor, you must first remove the fan. The housing holding the blades is usually held on the motor shaft by Allen screws. Loosen these screws with an Allen wrench and slide the blades off. Now the motor is accessible for repair or cleaning.

If the switch in a fan is at fault, remove it and replace it with a similar part. As an interim measure, you can bypass the switch until you get a replacement. The fan can be used in the meantime, but it will start when it is plugged in, and to stop it you must pull out the plug.

Some large attic fans have a belt drive between the motor and the fan blades. If the fan seems to lack power, it may be due to a slipping belt. There is usually an adjustable pulley for tightening the belt. If the belt itself is too worn, replace it with a new one.

A multiple speed fan usually has separate field windings, one for each speed. A selector switch connects the electricity to the proper winding for the chosen speed. If the fan runs on all but one speed, the trouble may be a bad contact in the switch or a burnt-out winding. In either case a continuity test will locate the trouble. Another way to determine the cause of the fault is to move the wires leading to the switch contact. For example, suppose the fan runs at low and high speeds, but won't run at medium. At the switch, swap the low and medium wires so that the low-speed field coil is connected to the medium position on the switch and vice versa. Now if the fan runs at its slow speed when the switch is at medium but fails to run when the switch is at low, then there is nothing the matter with the switch contacts. The trouble must be in the motor winding. On the other hand, if the motor runs at medium speed with the switch in low position, the medium contacts in the switch are at fault. If the switch is bad, it can be replaced easily. If a motor winding is at fault, the problem is more complicated. In some motors all the windings will have to be replaced since they are wound together. You must first take the motor apart before you can determine what new parts are necessary.

JUICERS

Electric juicers are very simple appliances. A reamer for squeezing the juice out of oranges, lemons, and similar fruit is mounted on the shaft of a motor. The motor is mounted so that the shaft is vertical and the reamer is at the top of the juicer. The motor itself is a simple split-phase motor which has low starting torque. This is why the motor will stop if you press down too hard with the orange you are squeezing.

Since the reamer should turn much slower than the motor, a gear box is mounted between the two to reduce motor speed to a suitable reamer speed. In some models, a strainer is added, so that the juice must pass through the strainer before coming out the spout. This strainer may be stationery, it may oscillate, or it may turn in the reverse direction. A reciprocating linkage causes oscillation, and an additional gear is used in models where the strainer rotates backwards. The gear box is usually fastened to the motor with Phillips head screws. At the other end of the motor shaft, a fan is sometimes attached to cool the motor when it is running. An illustration of the electrical circuit of a juicer with case removed is shown in Fig. 8-3. Note that the line cord is connected through a switch to the field coil in the motor. If a fan is included, as in the figure, the bottom of the case must have holes in it to let in air for cooling.

Because of its simplicity, a juicer rarely gives trouble. Line cord troubles are rare even after years of service. If a line cord is bad, it can be fixed by the techniques described in Chapter 3. The switch is the most likely candidate for trouble. If the juicer fails to operate when the switch is turned on, and if the line cord is plugged into a live receptacle, your first check should be continuity. Remove the plug from the receptacle and measure the continuity between the prongs of the plug, with the switch on and off. When the switch is off, of course, you should get an open circuit. When the switch is connected, there should be a continuous path through the switch and field coils, as can be seen in Fig. 8-3. If there is no continuity for either position of the switch, open the case and check directly across the switch and again directly across the field coils. The case is easy to remove, since it is held only by two to four Phillips head screws, and the motor and switch are easily accessible. If the switch is bad, replace it with a similar part.

SHAFT FOR REAMER

HUB FOR STRAINER

GEAR BOX

MOTOR

SWITCH

FIELD COIL

FAN

Fig. 8-3   Juicer circuit

If a motor winding is bad, you might still read continuity across the terminals, since the starting coil is also in the circuit. For motor troubles and their solutions, refer to Chapter 6. The motor in a juicer should give no trouble, however, unless liquid gets into it. Thus, never wash the motor casing, but simply wipe it off with a damp rag or sponge. If through neglect, the case is badly encrusted with fruit juice or other dirt, it can be removed from the motor and washed in warm water and detergent. If the case is plastic, as many are, it should not be washed in an automatic washing machine.

MIXERS

In an electric mixer, a motor drives two or three beaters for mixing, stirring, or whipping a variety of ingredients in the preparation of food. When mixing heavy batters, the beaters must turn slowly, and the motor is heavily loaded. On the other hand, when whipping cream or beating eggs, the load is light, and the motor rotates rapidly. Because the motor must operate under widely varying load conditions, including starting under heavy load, universal motors are used in mixers.

To meet the different speed requirements, some method of varying the speed is necessary. Three different methods of varying speed are described beginning on page 116 in Chapter 6. All three are used in mixers. In general, if the mixer has three distinct speeds which can be selected by moving a switch, the mixer uses a tapped field coil for speed control. If the speed is continuously variable, then the mixer uses either a movable brush or a centrifugal governor.

Thus, the motor in a mixer is versatile. It starts and runs under all sorts of load conditions, and its speed is adjustable. Why not then use it for many other uses as well as mixing? All that would be required would be to remove the beaters and substitute some other tool which can be motor-driven. Since this is indeed easily done, manufacturers do offer an assortment of attachments with their mixers, and in most mixers these are easily placed in position to be driven by the motor. These attachments include meat grinders, juicers, potato slicers, ice crushers, knife sharpeners, and many others. Despite the attractiveness of having a large group of functions in one appliance, many housewives prefer to use the mixer for its original purpose and consider it too much trouble to change attachments. This section will be limited to the mixer without attachments.

An outline of a typical mixer is shown in Fig. 8-4. In most modern mixers, the top part including the motor and beaters can be lifted off the stand so that the housewife can carry the mixing part of the appliance to her food, for example, to stir food while it is cooking on the stove. Notice in Fig. 8-4, that the brush holders are removable from the outside of the case. This is in recognition of the fact that, in a universal motor brushes wear and need to be

HANDLE

MOTOR
HOUSING

GEAR BOX

BRUSH
HOLDER

BEATERS
(2 OR 3)

LINE
CORD

STAND

Fig. 8-4   Mixer

changed. Thus, it is seldom necessary to open up the case, since brushes can be removed without opening it, and other motor troubles rarely occur.

The motor is mounted with its axis horizontal, and a worm gear mounted on the shaft engages other gears in the gear box to drive the beaters. Some mixers also have a shaft at the top of the gear box for attaching other appliances. The gear box is lubricated with a special grease which is sealed in and should last indefinitely. It is not necessary to oil the motor or other parts of a mixer with normal use.

Most mixers use capacitors to minimize radio interference. Also, those that use a centrifugal speed control have additional capacitors to prevent arcing at the contacts. In some models these capacitors are across the line voltage and grounded through another capacitor at their junction, as shown in Fig. 8-5. The ground in this instance is the case of the mixer. If the case is not actually tied to a good ground, there is a voltage of about half the line voltage (or

LINE VOLTAGE

(GROUND)

Fig. 8-5   Capacitors across line

about 60 volts) between the case and ground. When the plug is inserted in the receptacle, the case will be grounded through the ground line for one position of the plug, but will be floating if the plug is reversed. To avoid electric shocks, check for a voltage between the case and ground with the mixer both on and off. If there is a voltage there, reverse the plug in the receptacle, and that should remove the voltage on the case.

The most common trouble occurring in mixers is bent beaters due to carelessness or abuse. If the beater blades are bent, you can straighten them, but if the shaft of the beater is bent where it enters the mixer, it may be impossible to make it straight enough to enter without forcing it. You can buy new beaters to replace the old ones.

If, with the plug in a live receptacle, the mixer will not run, check the line cord and switch, making appropriate continuity tests. Then check the brushes. If they are stuck or worn, the motor will not run. To repair the unit replace the worn brushes. If the brushes stick, you will have to take apart the case and motor to get at the commutator. Sandpaper the brushes and commutator as explained in Chapter 6.

Other things which can prevent the mixer from operating are an open centrifugal switch on the speed control, an open coil in the field or armature of the motor, frozen gears, or encrusted food. Check the centrifugal switch. It

(a) TWO SQUARE SPINDLES   (b)TWO ROUND SPINDLES
                                 WITH EARS

(c) THREE SPINDLES

Fig. 8-6   Spindle arrangements

should move easily. It may be stuck in the open position due to lack of oil at the pivot point or too much dirt there. Clean and oil it as required. If the contacts are too worn, or the spring is broken, make the appropriate replacement.

An open coil can be detected by a continuity check. This and other motor troubles are described and remedies are presented in Chapter 6.

If the gears are frozen, it will be necessary to take the gear box apart. Clean the individual parts and repack the gear box with an approved lubricant according to the manufacturer's instructions. If this seems like too tedious or difficult a task, simply replace the whole gear box. Note that if the gear box is taken apart, you must observe carefully how each part fits. It is worthwhile jotting down the positions of all the parts on a piece of paper as you go along. When the gears are replaced, remember that the beaters must not interfere with one another. The spindle openings or driving slots must be in one of the arrangements shown in Fig. 8-6.

If the mixer is encrusted with batter or dirt, it should be taken apart and cleaned. The motor windings must not get wet, but other parts can be soaked in hot water and detergent. Use an old toothbrush to reach hard-to-get-at areas. After cleaning and drying, oil the bearings and reassemble.

If the motor sparks or does not run smoothly, the trouble is either in the brushes or commutator. Check the brushes and replace them if necessary. If the motor must be disassembled to get at the commutator, it is a good idea to clean it at the same time. Commutator and brush troubles and repairs are described in Chapter 6.

If the mixer runs but is noisy, it may need lubrication. Check the gear box and lubricate it if required. Other causes of noise are worn bearings or gears and an erratic centrifugal speed control, or governor. Replace defective parts as required. Excessive vibration can also be caused by an erratic governor.

If the motor runs, but the beaters do not turn, the fault may be broken gears or a broken or loose coupling. The pin holding the gear to the shaft may be sheared. Replace all defective parts and make sure new gears are properly lubricated.

If the mixer has no power, it may be dirty or in need of lubrication. Also check contacts in the speed regulator. Clean and lubricate them as required.

If the mixer speed cannot be varied, the capacitor across the contacts in the speed regulator may be shorted. This in effect shorts out the contacts so that the motor is never turned off part of the time, as it should be when the governor is operating properly. Remove the capacitor, and if the mixer now operates correctly, the capacitor is at fault and should be replaced. Note that the capacitor is necessary to prevent too much arcing at the contacts. The mixer speed will be constant if the governor contacts are stuck together. Replace the contacts if necessary, but sometimes sandpapering them and oiling the pivot arm will solve the problem.

In some centrifugal governors, a resistor is added across the contacts along with the capacitor to smooth out the running of the motor. This is shown in Fig. 8-7. Now when the contacts are open, as shown in the figure, the

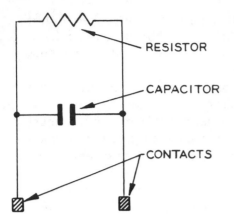

RESISTOR

CAPACITOR

CONTACTS

Fig. 8-7 Resistor added across contacts

resistor presents a path for current, and the motor is not shut off completely. However, because it is a high resistance path, the motor will run more slowly. Thus, the effect of the switch is to change the motor from full torque to a low torque, instead of from full torque to zero, as it would do without the resistor. This makes for smoother operation. If this resistor opens, the motor may be jerky. In that case, a continuity check across the resistor, with the contacts open, will reveal the trouble. Replace the defective resistor with a new one.

If grease leaks out of the beater sockets, it is a sign of worn bearings or thinned grease. Clean out gear box and fill with approved lubricant. If trouble persists, a new gear box will be needed.

BLENDERS

A blender is similar in principle to a mixer in that a motor drives a tool which acts on the food. In a blender the motor is below the mixing bowl, while in a mixer it is above. Both use universal motors. In general, a blender uses a more powerful motor because chopping and shredding requires more work than stirring and mixing. Heavy duty blenders draw 750 watts of power, but lighter, inexpensive models use as little as 250 watts.

In a blender the cutting knives are permanently mounted in the bottom of the bowl, which may be detached from the base containing the motor, as shown in Fig. 8-8. The bowl is held in position on the base by four ears or some similar locating device, and the motor shaft is then coupled directly to the cutting knives. The motor speed is controlled by depressing one of a set of push-button switches which connects the line voltage to a tapped field coil. By utilizing only a selected part of the

Blenders

BOWL

KNIVES

EARS

BASE

PUSHBUTTONS

Fig. 8-8   Blender

coil for each switch, the desired speed is obtained. Some older models have a single switch and run at only one speed.

In general, there are two ways to remove the case of a blender to reach the motor for testing and servicing. In some models, a hexagonal nut on the top of the base holds the motor, as shown in Fig. 8-9. You can unscrew this nut with large pliers or a wrench, and the motor can then be removed from the bottom. In newer models, you must turn the base over. The bottom of the base is held on by the same screws that hold the rubber feet in place and by two nuts holding the plate to the rotor. These are shown in Fig. 8-10. Remove all four feet and the two small nuts and then you can remove the bottom plate, exposing the motor.

The motor is usually surrounded by a heavy paper cone to protect it from dust and food particles. This should be removed carefully to get at the brushes and other parts of the motor. If the paper cone is torn in removal, you should replace it with a new one when reassembling the base.

If a blender will not run or runs without power when it is switched on, the first thing you must do is determine whether the trouble is in the base or the bowl. Remove the bowl and try again. If the motor runs well without the bowl, the trouble is not in the motor. You may be overloading the motor by putting too much food or items which are too hard to cut in the bowl. Try it with the bowl empty. If the knives still do not turn easily, the trouble is in the bushing. This problem is frequently caused by improper cleaning. Soak the bowl in a mixture of detergent and hot water until you can turn the knives easily with your hand.

The motor itself rarely develops trouble, although brushes may wear. For motor troubles and their cures, refer to Chapter 6. For line cord problems, see Chapter 3.

If the motor runs correctly at some speeds but fails to

Fig. 8-9   Top view of base

run at others, either the selected switch is open or the appropriate part of the field coils is burnt out. Switch trouble is more common. Check each switch by a continuity check, making sure nothing else is in the circuit, and replace any defective switch.

If the motor is taken apart, to check the commutator for example, make sure that all shims on the armature are replaced in the same positions they occupied originally.

ELECTRIC KNIVES

An electric knife simplifies carving of meat and poultry. The cutting action is performed by two knife blades arranged in a reciprocating fashion so that they are always moving in opposite directions. The blades remain in contact with each other so that a single cut is made in the food. The blades are removable.

The arrangement of internal parts is indicated in Fig. 8-11. The motor is a DC motor, which means it has

Electric Knives   **166**

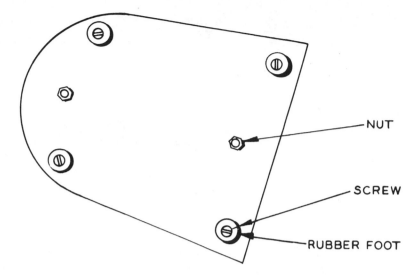

Fig. 8-10 Fasteners at bottom

brushes. A small rectifier converts the AC line voltage to a safe, low DC. The rectifier is a solid-state diode which should never give any trouble. The switch is a pushbutton or trigger on the case which is located conveniently at the finger tips so that it may be held in contact when the knife handle is grasped in the hand. Releasing the switch stops the cutting action. The motor shaft has a worm gear at the end of it, and this engages a pinion gear to which the knife blades are fastened. As the pinion gear rotates, the knives move back and forth in opposite directions. The motor shaft and worm gear are made of hardened steel and never wear out. The pinion gear, however, is usually made of fiber or plastic, so that the knife will run more quietly. The pinion gear does wear down eventually and needs replacing.

Possible troubles in an electric knife are:

1. the motor won't run

2. the motor runs, but the blades remain stationary

3. the knife operates too slowly or with reduced power

4. mechanical difficulties such as bent knife blades or broken case or handle

Electric Knives   **167**

Fig. 8-11   Electric knife

Some problems can be solved without taking the case apart, and as with all appliance troubles, these possibilities should be checked first. If it is necessary to open the case, you must remove several Phillips head screws first. The top of the case should lift off easily without force after these screws are removed. It is easy to overlook some of these screws, but all must be removed. With the top of the case off, it is possible to operate the knife and observe what is wrong.

If the motor won't run, check the line cord, outlet, and switch. Refer to Chapter 3 for line cord troubles and repairs. A continuity check across the switch will tell you whether the trouble is there. Contacts can be cleaned with sandpaper. If necessary, replace the switch. If voltage is getting to the motor but it still won't run, the trouble may be due to an open coil, a worn commutator, worn brushes, or bound gears. Check the brushes first since this is easiest. Refer to Chapter 6 for motor troubles and their correction. A continuity test across the motor terminals checks the coils, commutator, and brushes at once. Individual parts of the motor can be replaced as required. The most frequent motor troubles are worn brushes or a burnt-out commutator.

If the gears are bound, they may simply need grease or the pinion gear may be warped or broken. Special cream lubricant should be used because it doesn't run and leak out of the handle onto the food being sliced. A broken pinion gear can be replaced. In taking apart the knife, note carefully where each part is located and how it is fastened, and reassemble the parts in the same manner.

If the motor runs, but the blades don't move, the trouble is almost certainly in the pinion gear. Remove it and replace with an identical gear, adding a small amount of cream lubricant as needed.

If the knife is sluggish, the trouble can be due to worn brushes or dry bearings. Check the brushes for wear and also see that the stranded pigtails to the brushes are not broken. If necessary, replace the brushes. The bearings in these motors are made of powdered metal and are oil impregnated so that they should give trouble-free service. If they eventually do dry up, put *one drop only* of oil on each bearing. This should be sufficient for more than a year's use.

Mechanical problems are obvious. If the case is broken or cracked, it can be fixed with a suitable plastic cement. Bent knife blades should be replaced since they can never be accurately straightened.

Besides the standard electric knife which operates from the house line, manufacturers offer a cordless model. This is battery operated. When it is not in use, the knife is plugged into a battery charger which is always connected to the power line. Thus, the knife is always ready to operate, and on a single charge will run for more than enough time to do all the carving at a meal. The battery-operated model has five nickel-cadmium batteries in the handle to drive the motor, but is otherwise identical with the standard model.

The possible troubles in a cordless knife are much the same as those in the standard model with the addition of battery and charger problems. When the batteries are fully charged, you should measure about 7 volts across them without a load, that is, with the knife off. When the knife is running the battery voltage drops to something between 5.5 and 6 volts. If the battery voltage under load is much below 5.5 volts, one of the cells may be dead or there may be a fault in the charger. Measure the voltage across each cell separately with the motor running. You should read at least one volt across each cell. If one cell is dead, replace all of them.

The output of the charger should be about nine volts. If in checking the charger, you find no voltage or a low voltage, the trouble may be in the line cord or in the charger itself. Check the line cord to make sure voltage is getting to the charger. If the charger is at fault, replace it.

TOOTHBRUSHES

Like the cordless electric knife, an electric toothbrush also runs on rechargeable batteries. A small DC motor drives the brush in a rapid up-down motion to brush the teeth. The motor is completely sealed in a plastic case to ensure that no moisture enters to damage it. This means that if anything goes wrong with the motor, you cannot repair it. You can, however, send it back to the factory, where the motor will be repaired and sealed in a new case for a small fee.

When not in use, the toothbrush holder is placed in the charger which is always left connected to the house power line. The charger develops about 1 1/4 volts of direct current.

Troubles that can occur are the usual line cord ailments, a bad switch in the charger, or a defect in the motor. Also, if you drop the case, it may crack. You can check for line voltage at the input to the charger with a meter or test lamp. No voltage would indicate a bad switch or line cord, which you can fix. If there is voltage at the input, but no DC voltage at the output of the charger, you need a new charger. Replacement chargers may be obtained separately without buying a whole new electric toothbrush. If the charger is in working order, but the toothbrush motor refuses to run, you should get a new motor unit, which also can be obtained separately. If the motor runs, but the toothbrush won't move, the trouble indicated is in the gear or coupling. Again, this is something you cannot fix since the unit is sealed. It must be sent back to the factory for repair.

LINE CORD

HOSE

BONNET

Fig. 8-12 Hair dryer

A cracked case can be fixed with plastic cement. However, make sure that the unit is still operative since a fall which breaks the case can also damage internal parts.

HAIR DRYERS

The operation of a hair dryer is very similar to that of a space heater. In both, a heating unit heats the air and a motor turns a fan to blow the warm air where it is wanted. A space heater sends the warm air into the room, while a hair dryer directs the warm air to the damp hair which is to be dried.

The parts of a hair dryer are shown in Fig. 8-12. The motor is usually a shaded-pole motor which runs on AC only. This type of motor has low starting torque and low power, but since it only works against air, not much power is needed. The shaded-pole motor is relatively inexpensive and rarely gives trouble. The motor turns a fan blade, or impeller, which may be made of metal or plastic. It is usually held on the motor shaft with an Allen head screw.

In most popular models there are two heating elements, typically rated at about 100 and 200 watts. A switch connects either or both across the power line. Thus, for high heat, both heaters are connected; for medium, the

Fig. 8-13   Circuit of hair dryer

200-watt unit is used; and for low, the 100-watt alone. The motor is usually connected in the circuit in such a manner that it is impossible to switch on a heating element unless the fan is also running. In most models it is also possible to run the fan alone without a heater turned on so that cool air is supplied.

If the motor had a defect so that it failed to run when it was turned on, then it might be possible to turn on a heating element without having the fan running. This could be potentially dangerous, since the element would be heating the air inside the unit but the hot air would not be removed. The internal heat could increase to a point where it might cause permanent damage. To prevent this, most hair dryers have a safety thermostat in series with the power line to disconnect the voltage when the internal temperature gets too high. A diagram of the electric circuit of the dryer is shown in Fig. 8-13. The thermostat is normally closed. Note that the motor switch and heat selector switch are shown as separate entities, and that the motor must be turned on before voltage can be applied to the heaters. In practice both functions are combined in one rotary switch with five positions. These are:

1. motor and heaters *off*

2. motor on, heaters off, for cool air

3. motor on, low heater on, for low heat

4. motor on, high heater on, for medium heat

5. motor and both heaters on, for high heat

If an electric hair dryer is completely inoperative, that is, if the fan motor won't run and the heaters do not get hot, there is probably no voltage reaching the unit, since it is unlikely that both the motor and the heaters would fail at the same time. Assuming that the dryer is plugged into a live receptacle, the trouble could be in the line cord, switch, or safety thermostat. Check the line cord, using the techniques described in Chapter 3, and repair or replace it if necessary. If the cord is in order, next make a continuity check across the thermostat. It should look like a short circuit when cold. *Make sure the dryer is unplugged when you make these tests.* If the thermostat is defective, replace it with one just like it. Note that if the thermostat sticks in the closed position, there would be no indication to the user, since the hair dryer would operate as it should. However, it would no longer have the safety feature. Consequently, if the case of a hair dryer seems to get too hot when the dryer has run for a long time, you should check the thermostat to make sure the contacts are not welded together.

If the cord and thermostat are both all right, but the dryer shows no sign of life, the trouble is probably in the switch. You can use a clip lead across the switch to short it out of the circuit, and then the dryer should operate. This would show that the switch is bad and should be replaced. You can check in a more sophisticated manner by measuring voltages at the switch terminals. If the switch is bad and you decide to replace it, make sure you label all wires leading to it so that you can correct them to the corresponding terminals of the replacement switch.

For other troubles, you must first determine whether the fault is in the fan motor or a heater. This is fairly simple. If the heaters light, but the motor does not run, the trouble is in the motor circuit. It may be a fault in the motor or fan, or it may be a bad contact in the switch. Check the switch first. You can test to see whether there is voltage on the motor side of the switch when the switch is on. If not, the switch is at fault. If there is voltage reaching the motor and it still won't run, the trouble may be internal, such as an open coil or a frozen bearing. For motor troubles, refer to Chapter 6. A shaded-pole motor should give no trouble, but if it does have an open coil, it may be simpler to replace the motor rather than trying to fix the coil. If a bearing is stuck, it can be cleaned and oiled. If you take a motor apart to clean the bearings, make sure to note how each part, including shims, is placed, and reassemble in the same manner.

It is possible for a bobby pin or hairpin to come loose and work its way down the hose to the fan assembly, where it can stop the motor. Look for reasons why the fan might be stuck. You should be able to turn the fan freely with your fingers. Noisy operation may be due to the fan hitting its housing or loose impediments. This is fairly obvious and simple to correct. If the fan is bent, straighten it. If it hits the housing, loosen the Allen head screw holding the fan, and move the fan so that it no longer makes contact. Don't forget to tighten the screw.

If the motor runs sluggishly, check the bearings. Clean and oil them as needed. Make sure there is at least 0.005 inch endplay in the rotor.

If the motor runs, but neither heater lights, the trouble is probably in the switch, since both heaters should not fail simultaneously. If only one heater fails to operate, the trouble is probably an open heater, but it can also be a poor switch contact. In either case, continuity tests will enable you to locate the defective part. Switch contacts can be cleaned with sandpaper, but it is usually better to replace the switch. A bad heater should be replaced with a new one. Observe carefully how the wires are connected, making a sketch if possible, so that you will fasten the wires to the new part on corresponding terminals.

The air hose shown in Fig. 8-12 is made of a wire spiral covered with a tight-fitting plastic sheath. If someone steps on the hose or if it is otherwise abused, it can collapse or the plastic covering may be torn. The best repair is a new hose. However, since this always happens at night or on a weekend when stores are closed, you may have to make a temporary repair. Carefully push a broomstick through the hose and with your fingers bend the spiral wire to support the plastic. It may not look esthetically pleasing, but the hose can be made serviceable in this manner. If the plastic is cracked or torn, cover the damaged area with plastic electrician's tape.

CAN OPENERS

Electric can openers work exactly the same as the manual wall-mounted can openers except that the can is turned by an electric motor. An electric can opener uses a small shaded-pole motor to rotate the can under the cutter. Shaded-pole motors have very low torque, and you might suppose they would be incapable of exerting enough force to do the job. The secret is in the gearing. The motor shaft is coupled by a set of gears to a sprocket that turns the can. The gear ratios are such that the sprocket turning the can may rotate at a speed of 1/100 that of the motor shaft. A principle of physics states that the torque is then 100 times that of the motor itself, neglecting friction for the moment. Thus, because high speed is not needed, high torque is available through gearing from a low-torque motor.

In many models of electric can openers, a grindstone is attached to the motor shaft on the end opposite the gears that turn the can opener sprocket. This grindstone then rotates at the speed of the motor and can be used as a knife sharpener. A special slot in the housing of the appliance permits inserting a knife at just the right angle for sharpening on the grindstone.

An exploded view of an electric can opener with a knife sharpener is shown in Fig. 8-14. This is a Toastmaster Model 2210. Note that the motor rotor, part 28, is supported by brackets 25 and 27. The front end of the shaft of part 28 (toward the left in the drawing) has a small gear on it which engages a large gear, part 29. The

Fig. 8-14   Can opener with knife sharpener

center hub of part 29 engages the outside teeth of part 30. The center of gear 30 is connected to the drive sprocket 36. Through the double reduction gears, sprocket 36 turns much more slowly than the rotor and consequently exerts a correspondingly increased torque. The rear end of rotor 28 carries the grindstone 20 which is used to sharpen

Can Openers   **176** knives. Part 12 is a guide to hold the knife at the correct angle against the grindstone.

To use the can opener, the can is placed with its rim against the drive sprocket 36. When the handle, part 1, is depressed, it pushes the cutting wheel, part 39, into the top of the can and at the same time closes the switch, part 2. The magnet, part 42, is pressed against the top of the can to prevent it from falling inside when it is completely severed.

If the motor does not run when the switch is on, first check the line cord. Note that one side of the line cord is connected to a motor lead by a solderless connecter, part 8. This cap can be removed to check the voltage at the motor end of the line cord. Check the continuity of the switch also. If the switch is bad, replace it with a similar unit. The shaded-pole motor should give no trouble, although it may need occasional oiling at the bearings.

If the can opener is noisy, the gears or bearings or both need lubrication. To take the case apart, look for Phillips head screws around the case and remove them. The case should come apart without force, when all these screws are removed. Grease can be used on the gears, and light oil on the bearings. The gears will also be noisy if they are too worn. Worn gears should be replaced. It may be necessary to remove the motor to get at the gears, but you should not have to disconnect any wires. Notice carefully how parts are arranged, so that replacements may be reassembled in the same manner.

A common fault is that the can opener fails to cut all the way around the top of a can. This can happen if the carrier screw, part 37, is loose. However, a more likely reason is a worn drive sprocket or a dirty cutting wheel. Replace a worn sprocket if needed. If the cutting wheel fails to spin easily, it may be encrusted with food from months of service. It should be washed at least every six months and lubricated with light oil. If the cutter appears dull, you should replace it. Defective parts are easy to remove, but note carefully how they are positioned so that new parts are assembled in exactly the same way.

If the small gear teeth on the end of the rotor shaft are worn, you can replace the shaft. It is not necessary to replace the whole rotor. Before you remove the old shaft, however, measure how far the shaft protrudes from each end of the rotor. When you install a new shaft, make sure it protrudes the same amount from each end. Be careful not to lose any of the shims and spacers which are used to locate the rotor in the correct position in the stator. Look for them when removing the rotor and note their exact positions. When reassembling the motor, return these shims to their correct positions.

VACUUM CLEANERS

Vacuum cleaners may be small, hand-held, portable units or large industrial models. They come in a variety of shapes, usually variations of the three basic types shown in Fig. 8-15. The upright model, shown in (a), is self-contained, while the tank and canister types have separate hoses and a set of attachments for doing a variety of housecleaning tasks.

The upright cleaner has a rotating brush in the base which agitates the rug being cleaned so that the dust is released and more easily sucked up into the bag. The brush also helps pick up hair, thread, pins, and other small objects on the rug. The nozzle at the bottom of the upright vacuum cleaner is fairly large since it must be wide enough to hold the brush. In effect, this tends to reduce the suction power of the motor, since it is distributed over a wider area. Nevertheless, because of the brush, the upright is the best model for cleaning rugs. The brush is driven by a belt coupling it to the motor.

The tank and canister types do not have a brush, and manufacturers of these types try to compensate for the lack by using a more powerful motor and a smaller nozzle. Although inferior for rug cleaning, these models do a better job on hard floors such as wood, linoleum, or tile. They can also perform tasks not possible with the

(b) TANK

(a) UPRIGHT

(c) CANISTER

Fig. 8-15   Types of vacuum cleaners

upright, such as cleaning walls, drapes, or furniture. Some-times both the tank and canister vacuum cleaners are referred to as tank types, with the model shown in Fig. 8-15 (b) called a horizontal tank, and that in (c) a vertical tank. Either type may have skids or wheels.

All of the vacuum cleaners use the same principle of operation and essentially consist of two systems, one elec-trical and the other pneumatic. The electrical system con-sists of a universal motor, a line cord, and a switch. The uprights use a smaller motor, sometimes as small as 1/4 horsepower, while the tank-type motors may be 3/4 horsepower for home use, and more in industrial vacuum cleaners. In uprights the switch is usually a sliding device on the handle. In tanks and canisters, the switch is usually foot-operated or a toggle switch, but occasionally slide switches are used here also.

The pneumatic or air system sucks air in from the outside, passes the air through a bag to collect the dirt, and moves the air outside again. What is required then, is a *continuous* air passage through the bag. This means the bag must be porous so that air can go through, but of a fine enough mesh to trap the dirt. The air path is slightly different for the uprights and the tanks. In Fig. 8-16, the air path for an upright vacuum cleaner is illustrated. A large fan attached to the motor sucks air and dirt into the air intake. The intake is, of course, the nozzle which is pushed over a dirty surface. The dirty air is blown back through the outlet into the porous bag on the outside of the upright. The dirt is trapped in the bag, but the air passes through the pores in the bag back to the outside. The direction of air motion is indicated by arrows in Fig. 8-16. Note that air does not pass through the motor but flows around it.

The air path in a tank or canister vacuum cleaner is shown in Fig. 8-17. Here the bag is positioned at the air intake. Dirty air is sucked into the bag. Again the dirt is retained; but the air continues through the porous bag, past the motor, and back to the outside.

In the upright model, the bag must be emptied periodically and eventually loses its porosity when dirt cakes in the holes. In the tank types, the bags are usually discarded when they are full of dirt and are replaced by new ones which are quite inexpensive. In both types, the motor is usually protected from the air flow by a paper or cloth shield to prevent dirt from getting into the motor. Note that in Fig. 8-17, if the bag functioned perfectly to filter out the dirt, only clean air would reach the motor. However, some dirt does leak through the pores in the bag, and consequently the motor does need protection.

Troubles can arise either in the electrical or air systems, but you can easily determine where the fault is.

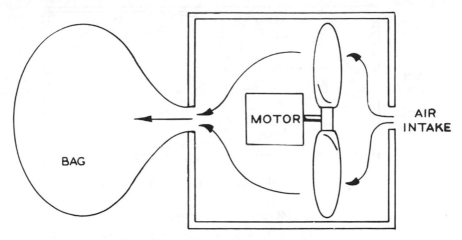

Fig. 8-16 Air path in upright vacuum cleaner

If the motor fails to run, the problem is in the electrical system. If it runs smoothly, but there is no suction, look to the air passages. If the motor runs sluggishly, however, it may be a fault in the motor, or it may be caused by a block in the air passage making it more difficult for the fan to pull air through.

Whenever there is a sign of trouble, try running the motor with the air passages open. In an upright, remove the bag and lift the nozzle off the floor. In a tank or canister, open the case and remove the bag so that the motor and fan are free. In either case, the motor should run easily when the switch is turned on. When the switch is moved to the off position, the motor should coast to a stop gradually. If it stops too abruptly, the cause may be worn brushes or dry bearings. In the case of the upright vacuum cleaner, a tight belt to the rotary brush or lint around the brush bearing can also impede the motor. In rare cases, dirt reaching the motor after many years of service may cause the motor to labor. For motor troubles and their cures refer to Chapter 6. If no voltage is reaching the motor, check the line cord and switch. Make sure the cord is plugged into a live outlet. For remedies to line cord problems refer to Chapter 3. Continuity checks across the switch will reveal a fault there. Replace a bad switch with a similar unit.

Vacuum Cleaners   **181**

Fig. 8-17   Air path in tank vacuum cleaner

The most common complaint is that although the motor is running, the vacuum cleaner does not clean as it did when it was new. The uprights differ from the others in possible causes for this trouble. For the uprights first check the rotary brush. As the upright vacuum cleaner is used, this brush is constantly rubbing against a dirty rug and eventually gets worn down. After a while the bristles no longer reach the rug, so that, even though the brush is revolving as it should, the vacuum cleaner has lost the "beating" action which the brush is supposed to provide. To take care of this wear, the position of the brush is adjustable, and all it may need is to be moved closer to the bottom of the vacuum cleaner. If the brush is completely worn, you can easily remove it and replace it with a new one. It is also possible that the brush is not turning because the belt driving it is loose or broken. Replace a frayed or broken belt. Tighten a loose belt by moving the brush.

If the brush and motor are in good shape, an upright vacuum cleaner can lose power from a clogged air system. Usually this means too much dirt in the bag, but as was pointed out above, eventually the pores of the bag get encrusted with dirt, and the bag needs replacing. Try running the vacuum cleaner without the bag, and if it works well, as indicated by a strong flow of air at the port where the bag is usually attached, look to the bag for the trouble. If there is little air pressure there, the fan might be loose on the motor, or there may be a wad of dirt inside the housing. Make the appropriate correction.

Vacuum Cleaners

182

Note that about the only possible location of a leak in the air system of an upright is at the point where the bag is attached to the housing. But a leak here will not prevent the vacuum cleaner from functioning. It will simply mean that some of the dirt is blown back into the room instead of passing into the bags. You can check for the possibility of a leak here by listening very closely or by observation when the cleaner is running.

Tank and canister types can have the same sorts of motor troubles or a loose fan blade, but no trouble from a rotary brush since there is none. If the complaint is poor suction, it may be due to a leak in the air system or a plug of dirt. In Fig. 8-17 only the body of the tank housing is shown. A long flexible hose is connected to the intake, and one of a variety of nozzles is attached to the hose. Leaks can occur at the junctions of the different parts and also in the flexible hose itself. Check the tank without the hose. When you place your hand over the intake hole, you should feel strong suction. If not, perhaps the bag is too full of dirt. The most common cause of poor suction is dirt clogging the hose. To correct this, attach the hose to the outlet port of the tank so that air will blow through it in the reverse direction. Usually the air will blow out the dirt and correct the difficulty. If the hose is broken so that leaks reduce the suction, buy a new hose.

Chapter 9

Large Appliances

Large electrical appliances are found in most house-holds and include refrigerators, dishwashers, clotheswashers and dryers, garbage disposers, and others. To be able to fix every ailment of every one of these appliances, the owner would need some special tools, as well as a large assortment of the more common household tools. When trouble arises calling for a special tool, it rarely pays to invest in one, since the same trouble may never arise again. To add to the problem, most of these large appliances are heavy or awkward to get at, and frequently an extra pair of hands, or some sort of moving apparatus is needed. Nevertheless, there are many faults you can correct, and if you can recognize these, you may be able to save the price of a few service calls.

UPPER
HOUSING

SHREDDER

Fig. 9-1   Upper housing of disposer

GARBAGE DISPOSERS

Garbage disposers grind or shred organic material into particles small enough to be washed down the drain. These materials include bones, vegetable matter including fibrous materials, and practically all food waste. Garbage disposers are *not* supposed to grind glass, string, bottle tops, silverware, or broken dishes. With proper use a garbage disposer gives trouble-free service for many years.

Because it must grind heavy bones as well as soft materials, a garbage disposer must be ruggedly built and be able to operate under varying load conditions. Usually it uses a capacitor-start motor which has a moderately good starting torque, although some less expensive models use split-phase motors. The motor turns an impeller and a flywheel, and there are no other moving parts.

DRIVING DISC

PIVOT HOLE

IMPELLER

FLYWHEEL

MOTOR

OUTLET

LOWER HOUSING

Fig. 9-2   Lower housing of disposer

The operation of a garbage disposer is illustrated in Figs. 9-1 and 9-2. Basically the machine consists of two main housings which are held together by bolts and nuts with a suitable gasket between the housings to prevent water leaks. The upper housing, as shown in Fig. 9-1, contains a chamber in which the garbage to be shredded is placed. At the bottom of the chamber there is a fixed shredder which is simply a metal ring with shredding teeth on its inside circumference. The bottom of the garbage chamber is open, so that the waste material rests on a disc at the top of the lower chamber. This disc is shown at the top of Fig. 9-2, which illustrates the bottom housing of the disposer. This housing contains the motor which rotates the driving disc and the flywheel. These parts are shown separated in the figure, but they are mounted on the motor shaft. Seals are required around the rotor shaft to prevent water from leaking into the motor. Washers, gaskets, and other hardware are not shown in the figures.

In operation, the motor rotates the disc causing the garbage to fly outward by centrifugal force. Near the periphery of the disc are two or more impellers, free to move on pivots, as shown in Fig. 9-2. These lie flat when the motor is at rest, but fly outward in motion so that they are close to the grinding teeth of the shredder. The garbage falls between the grinding teeth and the impellers and is shredded to bits. Cold water should be running while the disposer is in operation. This water washes the shredded garbage down the drain by way of the outlet shown in Fig. 9-2. A trap must be used in the outlet line. Note that cold water is used because it hardens any fat in the garbage, and the disposer is then able to grind the fat into tiny particles. If hot water were used, the fat would melt and flow into the drain where it could solidify when it cooled, causing a blockage.

A garbage disposer must have at least one switch to turn it on. Some units have a simple wall switch, similar to a light switch. Others have a switch which is controlled by the sink stopper so that it is impossible for the user to stick his hand in the disposer while it is on. An additional switch may be placed in the cold water line and is actuated only when the cold water is flowing. If this additional switch is used, both switches must be on before the disposer will operate. The motor also has a built-in overload switch which shuts off the electricity when the disposer has too heavy a load. In most models, the overload switch can be returned to the on position by pushing a button at the bottom of the unit.

Garbage disposers should give at least five years of trouble-free operation if used properly. Eventually the seals may crack letting water into the motor, or parts may wear out. However, occasionally a disposer may fail to operate because it is overloaded or stuck. This is easily corrected, and, in fact, is to be expected, since food particles sometimes lodge in such a position as to jam the machine.

If the disposer does not turn on, but the motor hums, it is usually a sign that particles of garbage have jammed the unit. This sometimes happens when the disposer has been shut off too soon leaving some food particles unshredded. Then the next time you try to use the disposer, these particles cause trouble by being wedged between the impellers or driving disc and the stationary part of the unit. With all switches off and with the disposer unplugged, if possible, reach into the opening and try to turn the driving disc. You can usually take hold of one of the impellers and use it as a handle to move the disc. *It should turn easily.* If food is jamming the disposer, the disc cannot be turned. Usually you can free the jam by turning the disc backwards. Manufacturers supply a special wrench, in the shape of a large Z, to do this. It is inserted in the opening and pushed against the impellers to free the food particles. You can also use a piece of wood. In some newer models, there is a special reversing switch, so that the disposer can be freed electrically.

If the jam cannot be cleared by using a wrench or piece of wood, it will be necessary to remove the bottom housing from the disposer. The upper housing is fastened to the sink. The bottom housing may be removed by unscrewing the six or more screws which hold the two housings together. This is not difficult but it might be awkward. There is not much room for a regular screwdriver, so it may be necessary to use an offset screwdriver. In addition, the bottom housing is heavy, and you must be careful not to let it drop when the screws are removed. When the two halves are separated, it is a simple matter to remove the particles of food or bone causing the obstruction.

Jamming can also be caused by frozen bearings. You can determine this when the lower housing is removed. It is possible to replace bearings for the flywheel, and on some disposers the motor bearings are also easily accessible. On others it is necessary to tear the whole motor apart. Fortunately bearings last many years, and when they do fail, it may be well to consider replacing the disposer rather than just the bearings. After many years the shredding teeth are worn, and the disposer doesn't operate as well as it did when it was new.

If the disposer shows no sign of life, make sure the overload button is reset and all switches are in order. Switches should be checked for continuity in the on position. *Make sure the disposer is disconnected from the power line when making these tests.* If the switch in the cold water line fails, you can bypass it temporarily. You must then remember to turn on the cold water when using the disposer.

The switch in the sink opening may show continuity but may be at fault because it has moved. Check to make sure that it is turned on when the stopper is placed in the proper position, and if not, adjust it so that it does and tighten the bolts that hold it in place.

If a disposer won't turn off, the trouble is a stuck switch. Repair or replace it as necessary. If the disposer runs and shuts off itself, it may be overloaded, or the overload switch may be defective. To replace a bad overload relay or switch, it is necessary to remove the lower housing and take out the motor.

If a disposer takes too long to grind the garbage, the impellers may be stuck so that they are not pushing the food against the shredder. Reach into the opening and try moving the impellers. You should be able to move them freely with your fingers. If one is stuck, you can pry it loose with a small screwdriver or the back of a teaspoon. On rare occasions you might have to remove the lower housing to remove the food particles. If the impellers are free, slow shredding is probably due to worn teeth in the shredder. Replace the shredder. To do this remove the lower housing and you can then reach the shredder in the top.

If a disposer leaks, this is usually a sign of worn or broken bearings. These can be replaced, but as a rule this doesn't happen until the disposer is old, and it is usually better to replace the whole disposer at this time.

REFRIGERATORS

All substances exist in one of three physical states, namely, *solid, liquid* or *gas.* It is possible for a substance to change its state by the addition or removal of heat from the substance. When a substance changes from solid to liquid, from liquid to gas, or directly from solid to gas, heat *must be added* to the substance. When the change of state is in the other direction, that is, from gas to liquid, from liquid to solid, or from gas to solid, heat must be *removed* from the substance. In practice, this means that, for example, when a liquid evaporates (changing from liquid to gaseous state), it must absorb heat from the surrounding medium. As an experiment, try moistening your hand with alcohol. As the alcohol evaporates, it draws heat from your skin, and your hand feels cool.

Anything which accelerates the change of state increases the addition or removal of heat. Thus, in the alcohol experiment above, if you fan your hand or blow on it, you will cause the alcohol to evaporate more rapidly, and your hand will feel cooler than it would without the additional air motion. Another method of increasing evaporation is by lowering the pressure. Water boils at 212°F at sea level, but on top of a high mountain, the boiling point will be several degrees less. The air pressure on the surface of the liquid is less on the mountain top, so that boiling takes place at a lower temperature. In a vacuum, water would boil even at its freezing point of 32°F. Below the boiling point, evaporation is increased as the pressure on the surface of the liquid is decreased, so that on a mountain top water evaporates more rapidly than it does at sea level. Conversely, if the pressure is increased, evaporation is slowed.

These two principles are the basis of operation of a refrigerator. Before electric refrigerators were invented, people used iceboxes. A piece of ice placed in the ice chamber slowly melted. The change of state from solid water to liquid water required heat, and this heat was drawn from the surrounding air and from the food in the icebox. If the ice were wrapped in paper to make it last longer, the rate of melting would be decreased, and the food would not get as cool.

In electric refrigerators, a special *refrigerant* is used. This is a substance which boils at a low temperature, much

Fig. 9-3   Simple refrigerator

below the freezing temperature of water. Another desirable characteristic of a good refrigerant is that it must change to liquid when the pressure is increased. There are many substances which can be used as refrigerants, all with long chemical names. They are usually referred to by their trade names. Typically, *Freon* boils at 22° below zero on the Fahrenheit scale, and *Refrigerant-22* boils below 40° below zero. In comparison, the boiling point of water is 212° F, and its *freezing point* is 32° F. Thus, Freon, for example, boils at 54° less than the temperature at which water freezes.

A simple refrigerator could be made as shown in Fig. 9-3. A jug of refrigerant is placed in one part of an insulated box with the opening of the jug leading to the outside. As the refrigerant boils or evaporates, it draws heat from the rest of the box, cooling the food there. You should recognize that this is essentially the same as the old icebox, with a liquid refrigerant replacing the block of ice. Both depend on a change of state for cooling the food. The refrigerator of Fig. 9-3 has several drawbacks, however. Refrigerants are expensive, and it would be necessary to keep replenishing the material that boiled away. Also, good chemical refrigerants may be harmful to health, and their vapors might be dangerous floating around the room. If a refrigerant like Freon is used, the temperature of the food would fall well below the freezing point of water, which is much too low for proper storage.

Refrigerators

Fig. 9-4   Basic circuit of refrigerator

A practical refrigerator then must have some means of catching the escaping vapor or gas and converting it back to liquid so that it is not necessary to add more refrigerant. Also, there must be some means of controlling the inside temperature by controlling the evaporation of the liquid. The basic circuit of a practical refrigerator is shown in Fig. 9-4. The refrigerant is contained in a closed system of pipes and is moved through the system by a compressor. The compressor is simply a pump driven by an electric motor. It sucks in warm vapor at the left in the figure and pushes it into a condenser, which is simply a length of pipe passing back and forth so that the whole length can be contained at the back of the case of the refrigerator. Note that the evaporator is inside the insulated box, but all the other components are outside, some at the bottom and others at the back of the case. Following the condenser the pipe leading to the evaporator in the insulated compartment has a restricted diameter so that the refrigerant between the compressor and the restricted tubing is under pressure. This increase in pressure condenses the vapor so that it becomes liquid again and in doing so gives off heat. The hot air rises from the coils of the condenser because it is lighter than the surrounding air, and thus cooler air moves in towards the condenser to remove more heat. In some refrigerators, this air movement is increased by the addition of a fan blowing across the coils of the condenser. From the condenser the liquid enters the restricted tubing and travels, under pressure from the compressor, to the evaporator. Here the coil diameter is increased so that the liquid entering the evaporator does so at reduced pressure. The pressure in the evaporator is further decreased by the compressor sucking out the vapor at the other end. The low pressure causes the refrigerant to evaporate quickly, drawing heat from the food stored in the refrigerator.

The temperature inside the refrigerator is controlled by a thermostat which is different from the bimetallic units used in most electrical appliances. The refrigerator thermostat has a bulb sealed to a bellows and containing a small amount of refrigerant. When the temperature increases, the refrigerant boils and causes the bellows to expand, closing a switch controlling the compressor motor. When the compressor has been running, the refrigerant in the sealed chamber contracts, and the bellows also contracts causing the switch to open, turning off the compressor. Provision is made for the switch to open and close with a snap action to prevent arcing at the contacts.

When the compressor is off, the pressure in the tubing equalizes quickly. Thus, when the compressor motor is started, there is very little load, and a motor with low starting torque can be used. The most common type is a split-phase motor. The motor and compressor are usually sealed in a single case with enough oil to run for 10 years or more. If trouble develops in this unit, the whole thing could be replaced, but the home handyman should not try to disassemble the unit. In general, by the time motor troubles develop, the refrigerator will have given many years of satisfactory service, and purchase of a new, improved model usually will prove to be more economical than trying to replace the motor.

Because of their size and weight, refrigerators do not lend themselves to easy servicing. Nevertheless, there are some things you can do, and you should not give up as soon as trouble appears. Fortunately, refrigerators last for a long time and when they do develop big troubles, the homeowner is ready to buy a new one. The repair of small problems is within your capability.

If the refrigerator fails to operate, and the light inside does not light, the trouble is almost certainly in the receptacle or line cord. Check the receptacle for voltage. Check the line cord for continuity. Sometimes a loose line cord connector falls out of the receptacle.

Refrigerator works, but interior light stays off. There are two possibilities here: either the bulb is burnt out, or the door switch is broken. Replace a burnt-out bulb, as required. If the switch is bad, replace it if you can get at it easily. On some models, however, replacing the switch requires much physical labor taking apart the box. You may decide to live with the bad switch, since the refrigerator works otherwise.

If the interior light works, but the refrigerator doesn't operate, the trouble may be due to a defective thermostat, bad contacts on the motor switch, or broken wires. Make a visual check for broken wires or thermostat troubles. Replace these parts as necessary. Some refrigerators use a capacitor-start motor, and the capacitor is easily accessible. Try spinning the motor with your hand while it is plugged in. If it then runs, the trouble could be the capacitor, a bad starting coil, or an open centrifugal switch. You can't do anything about the coil or other motor troubles, but you can replace the capacitor or switch.

If the refrigerator runs too long and the cabinet is warm, the door gaskets may be dried or broken. Put a piece of paper between the door and the frame and close the door. A dollar bill is the right size. Now pull on the piece of paper. There should be some resistance. If not, replace the door gasket. Another possibility for this trouble is poor insulation after many years of service. This is not worth fixing. The interior can also be warmed if the light does not go out when the door is closed. Check this by pushing the door switch. If the light stays on, replace the switch, if possible, or remove the bulb.

If the refrigerator runs too long and the cabinet is too cold, the problem is a defective thermostat, or switch contacts welded closed. Either part can be replaced.

If the refrigerator is noisy, look for the source of the noise. Loose condenser tubing may be vibrating against the frame. If the unit uses a fan to cool the condenser, check the fan blades for freedom and the fan bearings for wear. If the compressor is noisy, a possible source of the trouble is too much dirt around the condenser. Clean this area with a vacuum cleaner.

The operation of the refrigerator is affected by the weather. In hot weather, the unit will work harder to pull the temperature down, especially if the door is opened frequently. Operation is also affected by the amount of food inside. With a large quantity of food, a small amount of heat removed from each item can be sufficient to heat the refrigerant, and thus the refrigerator may not get cold enough. With an almost empty refrigerator, the required heat must come from fewer foods, and consequently the small amount of food may even be frozen.

Freezers and combination refrigerator-freezers work on much the same principles. The troubles you can fix are fixed in the same manner as those in refrigerators. Freezers should last indefinitely with no major troubles.

DISHWASHERS

In an automatic electric dishwasher, hot water and detergent are sprayed over dirty dishes with sufficient force to dislodge food particles. Dishes are then rinsed in clear, hot water. The process is repeated, and after several clear rinses, a Calrod heating element is turned on to warm the interior and dry the dishes. Once the dishwasher is started, each process in the cycle is controlled automatically by a timer of the sort discussed in Chapter 7 and illustrated in Fig. 7-7, on page 146.

Dishwashers may be built into the wall cabinets in the

Fig. 9-5   Impeller-type dishwasher

kitchen or may be portable, mounted on wheels, to be moved near the sink when needed and then wheeled out of the way. Some models, especially portables, are loaded from the top, but built-ins usually have a hinged door on the side. When a side door is used, there is always a door interlock, or switch, which shuts off the dishwasher whenever the door is opened.

The agitation of the water in the dishwasher is provided in one of two ways, either by an impeller or a spray arm. These two types of devices are illustrated in Figs. 9-5 and 9-6. Outside of the method of agitation the two types are similar. The motors are usually split-phase, although some models use shaded-pole motors.

In both types, the first operation in the washing process is draining the tub in case water was left from the last washing. All steps are controlled by the timer. The drain valve is opened, and the drain pump sucks water out of the washer and pushes it down the drain. In Fig. 9-5, a single motor and pump are shown, which is typical of impeller-type washers. In Fig. 9-6, two motors are shown, one for running the drain pump and one for running the

Dishwashers   **197**

Fig. 9-6. Spray-arm dishwasher

circulating pump. The two pumps are needed on spray-arm dishwashers and two separate motors are usually used, although one motor and a system of gears and clutches could be used. Draining lasts less than a minute.

After the drain operation, the drain valve is closed, and the water inlet valve opens. This latter valve is in the hot water line. Water should be between 140° and 160° for proper performance. Water enters near the top of the tub, although the inlet valve is usually located under the tub near the motor. During the fill operation, the motor or motors are usually at rest. The inlet valve remains open for a fixed time interval set by the timer. The amount of water that enters is a function of the water pressure, and if the pressure were too great, too much water would enter and possibly leak around the edge of the door. To prevent this, many models also have a float valve which is in series with the inlet valve and closes when the water reaches its maximum permissible level. No more water can enter even though the inlet valve stays open for the whole time interval set by the timer.

After the inlet valve closes, there should be about three quarts of hot water in the tub. Here the two types of washers differ slightly in operation. In the impeller type, shown in Fig. 9-5, the water level is above the top of the impeller. The motor is switched on and turns the impeller, which splashes water over and around the dishes with great force. Some of the water mixes with the detergent contained in an open detergent cup in the door of the washer. As this water falls back and is splashed out again, all the water gets mixed with detergent. After this wash operation, which lasts five or six minutes, the impeller stops, and the motor drives the pump to drain out the water through the drain valve, which is now open.

In the spray-arm dishwasher, the water level is below the spray arm. The recirculating pump drives the water up through the spray arm, and out openings in the arm. Because these openings are offset, the arm revolves and sprays all over the interior and mixes with detergent just as in the impeller type. At the end of five or six minutes the recirculating pump is shut off, and the drain pump starts and drains out the water.

Clean water enters and a rinsing operation takes place, since the detergent cup is now empty. This lasts three to five minutes.

After the rinse water is drained out, a second detergent cup opens, and another wash operation begins. This may last up to 15 or 16 minutes. The dirty water is removed, and two or more rinses with clear water follow. Finally, all the water is drained out, the motor is turned off, and the Calrod heater turns on to dry the dishes. If you open the door of the dishwasher soon after the end of the cycle, *do not touch* the Calrod heater, since it will still be hot.

A common complaint is that dishes are not cleaned thoroughly in an electric dishwasher, but more often than not this is caused by improper use. First you must consider the nature of the "dirt" on dirty dishes. Those foods that are soluble in water, such as sugar, syrups, fruit juices, and raw eggs are easily washed in the dishwasher. Also, oils and fats are melted by the hot water and emulsified with the detergent, so that these present no problems, either. Foods such as bread crumbs and gravies are not soluble, but the scrubbing action of the spray usually breaks up the particles and removes them from the dishes. Some foods, however, *if allowed to dry on the dishes,* are very difficult to remove either by hand or in a dishwasher. These include cooked eggs, tomato juice, apple sauce, and others. If these foods are served, the dishes should be soaked until they are ready to be put into the machine. Also, when placing dishes in the dishwasher, you must be sure that they face the direction of the spray, that they are not nesting together, and that the detergent cup is not blocked from the spray by a large dish.

Improper washing can also be caused by water being the wrong temperature. Make sure it is in the range of 140° to 160°.

If the washer takes too little water, it can also cause unsatisfactory cleaning. If too much water enters, it may leak out around the door. In either case the trouble may be in the timer or the intake valve. Check the valve solenoid for continuity. Check that the valve moves easily. Timer troubles and their corrections are discussed in Chapter 7.

If the drain valve leaks it may be due to a particle of food holding the valve open or a defective solenoid. Move the valve lever open and shut several times to clear the particle. Check the solenoid for continuity. A leaky inlet valve is caused by a bad solenoid or a defect in the valve itself. Both drain valves and intake valves are easily replaced if defective. Solenoids can be replaced without replacing the valve.

If the washer continues to operate when the door is opened, the interlock switch is defective. Clean the contacts, if possible, or replace the switch.

Noisy operation may be due to a broken impeller, interference, or improper stacking of the dishes. This is always a mechanical difficulty and locating the source of the noise usually pinpoints the solution.

If the washer will not start, assuming the power line is live, the trouble may be in the interlock switch or the timer. First check the switch and the timer motor for continuity. Finally make sure timer is not jammed, and replace defective parts.

If the main motor will not run, it may have a burnt out winding. On impeller-type machines, if the impeller is stuck because of improper loading of dishes, the motor also will not turn. Make sure impeller is free. Check the continuity at motor terminals. Motors should last for many years, but eventually a bearing leaks, and water gets in, ruining the motor. Replacing the motor is a big job because of its inaccessibility. It is necessary to pull out the dishwasher, after making sure it is unplugged, and tip it on its side to reach the motor. With help, it can be done.

If the dishes do not dry, the Calrod heater is not working. The heater itself may be defective or it may not be getting voltage. Check the heater for continuity. If it is open, replace it. If it is good, check the timer switch which turns on the heater. Also look for loose connections.

Make sure the cabinet is grounded to prevent electric shocks. It should be grounded when it is first installed, but if you pull it out for a major repair, you might break the ground wire. *Always check for safety after servicing.*

CLOTHESWASHERS

The process of washing clothes involves many separate and distinct operations. First the clothes should be soaked to loosen as much dirt as possible. The dirty soak water should be removed and fresh water added. Some sort of beating action now takes place, similar to scrubbing clothes on a washboard, to make sure the soap or detergent penetrates to all parts of the clothing. The water is removed, and this is followed by one or more rinses with clear water. After the last rinse, the water is removed, and the clothes are dried.

In an automatic clotheswasher, all these operations are performed in the proper order and at the proper time. The clothes do not come out completely dry, but the final operation is a rapid spin which removes most of the water by centrifugal force. Manufacturers have designed many different types of washers to perform these operations, but they fall into two categories: (1) washers with cylindrical drums, and (2) washers with agitators.

In the cylinder-type of washer, the clothes are placed in a cylindrical drum which rotates about a horizontal axis. The cabinet in which the cylinder is mounted is essentially a fixed tub, and the cylinder is perforated, so that when the tub is full of water, the water will surround the clothes inside the cylinder. A motor beneath the tub is coupled to the cylinder through a belt drive or set of gears. The access door to the cylinder is usually at the front of the cabinet, and the clothes are placed into the end of the cylinder, as depicted in Fig. 9-7 (a). This is called an end-loaded or front-loaded washer. In some cylinder-type machines, top loading is used, and the cylinder is closed at the ends but has a door on its circumference, as shown in Fig. 9-7 (b).

The agitator-type washer is shown in Fig. 9-8. A perforated basket mounted on a vertical axis replaces the cylindrical drum, and an agitator is added at the center. The motor, at the proper times, is coupled to the basket, to the agitator, or to the drain pump, as required.

In both types, a timer controls the operation. This is a clock motor with cams on its shaft as described in Chapter 7. First, the inlet valves are opened so that hot or cold water or the proper mixture can enter. After a predetermined time interval, the inlet valves are closed, and the clothes are "scrubbed". In the agitator-type machine, the agitator moves back and forth about its axis so that the clothes are tossed about in the water. In the cylinder-type, the scrubbing action is caused by baffles on the inside of the cylinder carrying the clothes around with them until they fall to the bottom as they near the top of the arc. The soaking and agitation may take 10 or 12 minutes. Then the water is spun out by rotating the basket or cylinder at high speed, and the pump empties the water out of the tub. The drain valve is opened for this operation and closes again when it is over. This is followed by more fill-agitate-spin operations. In some machines extra rinses

(a) END LOADING

(b) TOP LOADING

Fig. 9-7   Cylinder-type clotheswasher

are provided. Finally, the clothes are spun, and everything is turned off. The intake and drain valves are solenoid operated, as described in Chapter 7. The solenoids
are controlled by the timer, which also controls move-
ments of gears and belts, so that the motor drives the
proper pump, agitator, or basket.

When something goes wrong with a washing machine,
even if you diagnose the trouble, you may wonder how to
take the panels and top off so that you can make the
necessary repairs. Disassembly techniques vary from machine to machine. If there are no screws in view holding
down the top, slide the top toward the front and it can be
removed. The sides and front and back can then be
removed by taking out screws holding them. On agitator
machines, you must remove the agitator to get at the tub.
The top of the agitator can be unscrewed from the shaft,
and the agitator lifted out. If it sticks, loosen it by tapping
on the shaft lightly with a rubber mallet while exerting
upward pressure on the agitator. After the agitator is out,
the shaft can be removed by loosening some Allen head
screws near the base. After this, all disassembly is straight-

Clotheswashers

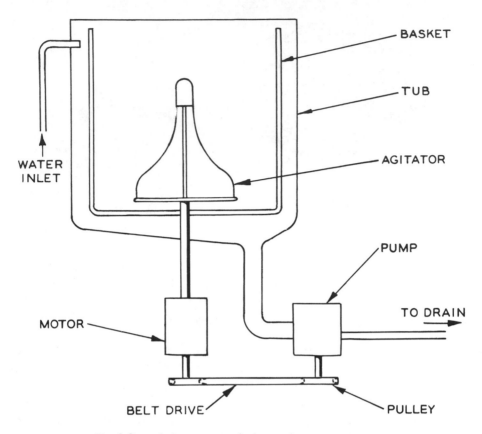

BASKET

TUB

AGITATOR

WATER INLET

PUMP

TO DRAIN

MOTOR

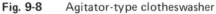

BELT DRIVE

PULLEY

Fig. 9-8   Agitator-type clotheswasher

forward. However, do keep a record of how things are taken apart so that you can reassemble them correctly. Important: when lifting off the part containing the timer, handle it carefully.

A common trouble in a washing machine is a jamming or blockage caused by a piece of clothing wandering out of the basket or cylinder and getting stuck. It can jam the works, stopping the machine, or can even deflect the water so that it spills on the floor. Always suspect a sock or handkerchief whenever the machine develops such trouble.

In most machines, hoses from the hot and cold taps are permanently connected, and these taps are left open. When the intake valves open, water flows through these hoses, but when the valves are closed, the water in the hoses is under pressure from the water mains. Presumably the hoses could rupture under this pressure, although it rarely happens. However, to prevent possible ruptures, especially when going on vacation, many users shut off the faucets that feed the washer, and then they forget to turn them back on and wonder why no water enters the machine. If no water enters, check the faucets. If the faucets are open, the trouble may be due to a bad solenoid on the inlet valve, a defective valve, or a fault in the timer. Tests for these faults and corrections are described in Chapter 7.

If the motor does not run, make sure voltage is getting to the motor. Check motor for continuity. The motor is usually a capacitor-start type, although split-phase motors are also used. Remove the load from the motor and check it by the techniques described in Chapter 6. If necessary, it can be replaced. However, washing machine motors usually outlast most other parts. If the motor finally fails, other parts of the washer may be on their last legs, and it may not be worth replacing the motor.

If the washer fails to perform one of the required operations in the washing cycle, check the timer, relays, and wiring. Look for loose or broken belts, or grease on clutches. Most machines have a switch which is actuated by the lid so that agitation stops when the cover is opened. If this switch is defective, the washer will not agitate. You can check the switch by a continuity test across it, with the washer unplugged.

If the washer is noisy or vibrates excessively, the clothes are probably distributed unevenly in the basket or drum. Redistribute the clothes.

If the washer leaks, check all gaskets, seals, and hose connections. Also check hoses for breaks. Replace or tighten as required.

It is most important that the washing machine be grounded, since it is used next to a good ground, the cold water pipe. After you have finished servicing it, make sure the ground wire is still connected. Also, as a precaution, measure the voltage between the case and ground both while the machine is running and when it is off. *The voltage must be zero.*

Fig. 9-9   Tumbling drum

CLOTHES DRYERS

In a clothes dryer, damp clothes taken from a washing machine are tossed about in a current of heated air until they are completely dry. Although dryers take many forms, all operate on much the same principles. The clothes are placed in a perforated drum with baffles, as shown in Fig. 9-9. As this drum rotates, the clothes are tumbled about.

The components of a clothes dryer are shown in Fig. 9-10. The drive motor is a split-phase type of about 1/3 horsepower, designed to run on 115 volts, AC. A large pulley on the drum is coupled to a pulley on the motor by a belt so that the drum rotates at about 50 rpm, much slower than the motor. Another belt couples the motor to a fan which sucks in air from outside and blows it over a heater and through the drum. The damp air is then blown out through a vent usually leading outside the building.

The source of heat in a dryer may be either gas or electricity. In a gas dryer, there must be some means of igniting the gas when it is turned on, and this is usually done by electricity, although in some models a pilot burner is on all the time. The main burner is located in the path of the air being blown toward the clothes. In an electric dryer, the heating element is usually located in a strip

Clothes Dryers   **206**

Fig. 9-10   Dryer components

around the entire perimeter surrounding the drum. The heater is a coil of nichrome wire which operates on 230 volts and puts out about 5000 watts. Thus, an electric dryer must be connected to a 230-volt line. The motor, however, runs on 115 volts, which is obtained between one side of the 230-volt line and ground, as shown in Fig. 1-6 on p. 8. A gas dryer, of course, must be connected to a 115-volt line to run the motor.

In a gas dryer, if the flame is blown out accidentally, it is important to close the gas supply, since gas fumes are dangerous. A special circuit is used to prevent gas

Fig. 9-11   Gas controls

flow when there is no flame, as shown in Fig. 9-11. When the dryer is turned on, current flows through a platinum heating coil (not shown in the figure) located just above the pilot jet. At the same time this current flows through the solenoid which opens the gas line to the pilot jet. Gas flows out of the jet and is ignited by the heating coil. A *thermocouple* is located above the pilot flame. A thermocouple is a device which produces a voltage when heated. The voltage produced in the thermocouple by the pilot flame actuates the solenoid valves to hold open the gas lines to the pilot and the main burner. The current to the heating coil is shut off, also removing the original current to the solenoid. The solenoid valve, however, is now held open by the thermocouple. If the flames blow out, the thermocouple would cool and no longer provide a voltage to actuate the solenoids. The solenoid valves would close, shutting off the supply of gas. The voltage developed in the thermocouple is much less than one volt, but is sufficient to hold the solenoid valves open.

When a dryer is turned on, the drum should start revolving immediately. Heat is not applied, however, until the drum is almost up to full speed. This is a safety precaution, since if the clothes did not tumble, the heat could be concentrated on one article of clothing, perhaps burning it. A centrifugal switch is used to start the heat *after*

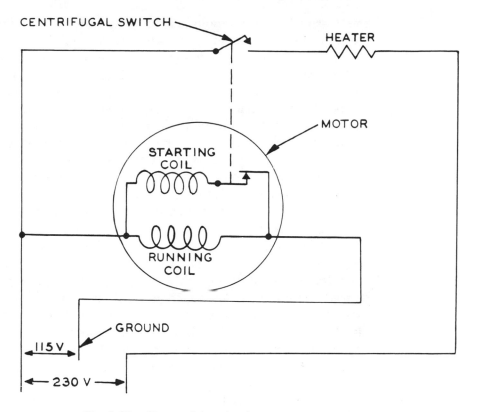

CENTRIFUGAL SWITCH

HEATER

MOTOR

STARTING COIL

RUNNING COIL

GROUND

115 V

230 V

Fig. 9-12 Heater delay circuit

the motor is rotating fast enough. The basic circuit in an electric dryer is shown in Fig. 9-12. A centrifugal switch is in series with the heater across the 230-volt line. This switch is mounted on the motor and is tied to the switch which opens the starting coil circuit. Thus, when the motor gets up to speed, centrifugal force opens the starting coil switch and closes the heater switch simultaneously. In a gas dryer, a similar centrifugal switch is used in series with the main gas solenoid. Thus, the solenoid valve does not open to permit gas flow to the main burner until the motor is up to speed. Note that it is possible for the motor to be running, but the drum to be stationary. This could happen, for example, if a belt broke. In this case, the heat would be turned on as soon as the motor got up to speed even though the clothes would not be tumbling. For this reason, the user should always make sure the clothes are tumbling when the machine is first turned on.

Some of the operations in a dryer are controlled by a timer, similar to that described in Chapter 7. The user turns a knob setting the length of time of operation, and as the dial turns back to zero, cams operate the proper circuits. The motor is turned on immediately, turning the drum and the blower fan. These remain in operation throughout the cycle. A switch actuated by the timer turns on the platinum heating element which ignites the gas in a gas dryer. After a suitable interval, the switch is opened again by the cam in the timer. About 10 minutes before the end of the drying cycle, the heat is shut off by the timer. This allows the clothes to cool somewhat so that they can be handled when the dryer stops.

Dryers are equipped with safety devices to prevent damage to clothes, to the machine itself, and harm to the user. The door has a switch which is in series with the main line, and shuts off the machine when the door is opened. In addition, thermostats prevent the interior from getting too hot by turning off the dryer when the air reaches an abnormally high temperature.

Most common faults in a dryer are mechanical. If the dryer motor runs, but the drum doesn't turn, the trouble could be a loose or broken belt or a pulley slipping on its shaft. Make sure the drum is free to turn. Sometimes an article of clothing can jam it. Then check belts and set-screws on the pulleys. Replace broken belts as required.

If the dryer operates, but the clothes remain damp, either there is no heat or the fan is not blowing air through the clothes. Check the fan belt first. Also check the fan bearings to make sure the blower is free to turn. If the air coming through is cold, the trouble could be in the heat source or switches. If the heater in an electric dryer is defective, it cannot be repaired, but you can replace it with a new one. In a gas dryer, the most frequent trouble in the gas line is failure of the solenoid valve which actuates the pilot. Another trouble is a burnt-out heating element for the pilot. In either case, replace with an identical part.

If the dryer won't run at all, first make sure the outlet is live. Line cords rarely give trouble, since they are never disconnected or moved. If there is voltage at the outlet, check the door switch, the timer, and finally the motor. Check all connections. Make continuity and voltage checks to ascertain if voltage is getting to the motor and timer. For motor troubles and their cures, consult Chapter 6.

If the clothes dry too slowly, the clothes may have been too wet when they were put in. This difficulty can also occur if the dryer is overloaded. Also, check for restriction of the air path. If the user fails to clean the lint trap, it may eventually get clogged and block the flow of air. A loose belt can also reduce the flow of air.

If the trouble is in the gas line, the trouble may be a solenoid valve, as mentioned earlier, or it may be a defective thermocouple. You can call the gas company, and in most areas, they will send a representative to determine the source of trouble. However, he will not fix it. If he identifies the fault as a defective thermocouple, you can replace it yourself.

A thermocouple externally looks like a long copper tube with a small bulb at one end. When heat is applied to the bulb, a voltage is produced at the other end. Usually the thermocouple can be removed with a small crescent wrench and replaced with a new one. Thermocouples are used in gas water heaters and gas furnaces, as well as gas dryers, and can also be readily replaced in these appliances when they are defective.

Chapter 10

Preventive Maintenance

The average home has about 20 to 30 electrical appliances. If each appliance had only one failure per year, you would expect then to have an average of a failure every other week. Clearly this is too many, but fortunately most appliances, even when abused, give good performance for much longer than a year. Nevertheless, when an appliance does fail, it is always inconvenient, and it is all the more painful because the failure probably could have been averted or at least delayed for many months by using proper operating procedures and suitable *preventive maintenance.*

If you have to call a repairman, you can expect to pay a bill which is large compared to the original cost of the appliance, even to replace a part costing only a few cents. In view of the high cost of repairs, it is truly amazing that more than one in ten service calls is to replace a burnt-out fuse, to tighten a loose fuse, or to plug in an appliance because the user didn't realize the plug had been pulled out. If you are able to do your own appliance repair work, you will solve the fuse problem or plug in the appliance in situations like these, and you can then congratulate yourself on how much money you've saved. It takes little effort, however, to make sure that the fuses (or interlocks) are good in the first place, and that line cords of frequently used appliances remain plugged in. By the same token, the amount of extra effort needed to operate appliances correctly and to keep them in operating condition is trifling compared to the extended period of trouble-free operation which results.

One of the greatest causes of failure in an electric appliance is not a fault of the appliance at all, but is a failure in the line cord or plug. After reading Chapter 3, you know what types of failure you might expect in line cords and plugs, and hopefully you now know how to fix them. However, almost all of these failures can be prevented! Here are some reasons why line cords fail:

1. The user yanks the plug out of the outlet by pulling on the cord.

2. The cord is bent sharply around an obstruction, or a piece of furniture is pushed against a plug, causing a sharp right angled bend in the cord.

3. The appliance is dangled by the cord.

4. The cord is run behind a radiator or along a window sill in bright sunlight.

5. The cord is run under a rug and is stepped on frequently.

6. Someone steps on a plug, cracking it.

This is by no means a complete list, but it should give you a good idea as to the causes of cord failure. Also, it is not difficult to figure out how most of these troubles can be prevented. The insidious part of this problem is that even when the cord is abused frequently, the appliance continues to work well, so it seems hard to justify the extra effort, for example, in reaching over to pull out the plug by grasping the plug itself instead of yanking the cord. Nevertheless, every time the cord is yanked it shortens the time to the next failure.

When you first buy a new appliance, read the instructions! As new improvements are designed into appliances, they also lead to new methods of operation. It is important for the user to know the limitations of an appliance as well as its advantages. Unfortunately, salesmen stress the advantages and gloss over the limitations. When you first buy an appliance, it is best to follow the manufacturer's instructions to the letter. However, auxiliary supplies are also improved, and some changes may be in order. For example, if your dishwasher requires one teaspoon of detergent (according to the instructions supplied with the machine), it may be desirable to use only 1/2 teaspoon of a new, improved detergent.

An electric blanket should give many years of service, when used according to the manufacturer's directions. The most common source of trouble in an electric blanket is overheating caused by leaving a weight on the blanket while it is on. The weight may be only a book which you intend to read in bed, but if it rests on the blanket too long, the heat under it cannot escape and may burn the insulation or the wire itself. Also, if an electric blanket is "tucked in", it is bunched or folded so that again heat cannot escape. It should be obvious that you should never sit on an electric blanket, even when it is off. The heavy weight on the fragile wires may cause them to break. In summary, electric blankets will give long life, if you remember these restrictions:

1. Don't sit on the blanket.

2. Don't put a weight of any kind on the blanket when it is on. This includes a bedspread over the blanket.

3. Don't tuck in the corners.

4. When the blanket is folded and put away, don't place any weight on top of it.

Where applicable, these same restrictions apply also to heating pads.

If a little is good, a lot is *not* necessarily better. This is an important rule to keep in mind when using detergents in dishwashers and clotheswashers. Too much detergent can harm an appliance, although the new low-suds detergents cause less trouble than the old sudsing varieties. If you have soft water in your pipes, you should probably use only half the amount of detergent that is recommended by the manufacturers. It is well to experiment when you first get your washer, and use as little detergent as can do a satisfactory job. Special rules for dishwashers are:

1. Make sure the water is at or above the minimum temperature specified by the manufacturer.

2. Don't use the dishwasher as a garbage disposer. Dishes should be scraped first.

3. Learn how to clean the filter and keep it clean.

4. Make sure silverware cannot interfere with the operation of the dishwasher.

Garbage disposers are real work savers, but are abused frequently. They cannot dispose of everything. It goes without saying that spoons, forks, and other items of silverware should not be dropped into the disposer. Items which are not satisfactorily ground include fibrous materials such as corn husks, watermelon rinds, and celery. Avoid raw bones. Although these would be ground up eventually, they put too heavy a load on the machine. If the disposer jams, don't panic. Use a long wooden stick to turn the turntable backwards and loosen whatever caused it to jam. You may have to press the restart button to start it again.

Toasters require some special attention. As bread is toasted, pieces frequently break off and fall inside the case. In most cases, small pieces and crumbs fall all the way through and may be removed through an access door. Larger pieces, however, can get caught in the body of the toaster and jam the mechanism or catch fire. It is important then to make sure the toaster is kept clean. The access door should be opened and all crumbs removed periodically. Once a month is not too often. Larger pieces should be removed whenever you know they are there. *Make sure the plug is disconnected whenever you clean out the toaster!* Also, make sure you plug it in again when you finish cleaning it.

A toaster probably takes more abuse than any other appliance. Typically, at breakfast the user is in a hurry or sleepy. She (or sometimes, he) puts bread into the toaster and pushes down the handle, but fails to push it all the way down. The bread pops up immediately. Now slightly irritated, she slams the handle down. This may bend vital parts, and although the toaster still works, its useful life is shortened.

Refrigerators and freezers should give many years of service. Some important steps to help achieve this are:

1. Make sure the condenser coils are kept clean by vacuuming at least every six months.

2. Do not keep the refrigerator or freezer in a warm place as it will have to work harder.

3. Don't pile things around or on top of the unit as it works better when air circulates freely around it.

4. When the weather is warm or when an extra amount of food is in the cabinet, turn up the coldness control.

5. Check the door gaskets periodically and change them before they become brittle with age.

Motors in many appliances are sealed, self-lubricating units and need no maintenance. In others they should be oiled once or twice a year. As a rule of thumb, if there is an oil cup on a motor or blower or other moving part, you should oil it. But don't flood it with oil. Two or three drops each time are sufficient.

Dirt is a bugaboo in all electric appliances. It was

already mentioned that the condenser on a refrigerator should be vacuumed periodically. Similarly, clothes dryers, air conditioners and other large appliances should be checked for dirt and lint. It is a good idea to remove the sides of the cabinet and vacuum out any dirt you can see. Also, filters should be cleaned or changed according to the manufacturer's instructions. For air conditioners and gas furnaces, filters are cheap and should be changed about once a month.

Vacuum cleaners are used to suck up dirt and dust, and consequently, it is surprising that they, too, must be kept clean. Usually, some lint or dirt remains in the hose when the vacuum cleaner is put away. After a few months the lint in the hose limits the amount of air that can enter. One way to prevent lint from remaining in the hose is to run the vacuum cleaner, sucking in clean air, for a few minutes after finishing each day's cleaning. This will cause the lint in the hose to be sucked into the bag. In some cleaners, it is possible to attach the hose to the blower end of the machine, and actually blow out the lint.

With some attention to preventive maintenance, appliances will last for many years. In fact, you may reach a point where you hope an appliance fails so that you can buy a new, improved model. Don't despair because your appliances last too long. If the new model is that attractive, you can donate the old one to a PTA rummage sale, and feel virtuous when you buy the new one.

Index